Lecture Notes in Mathematics

Edited by A. Dold and B. Eckmann

1143

Aloys Krieg

Modular Forms on Half-Spaces of Quaternions

Springer-Verlag
Berlin Heidelberg New York Tokyo

Author

Aloys Krieg
Mathematisches Institut der Westfälischen Wilhelms-Universität
Einsteinstr. 62, 4400 Münster, Federal Republic of Germany

Mathematics Subject Classification (1980): 10D20

ISBN 3-540-15679-8 Springer-Verlag Berlin Heidelberg New York Tokyo
ISBN 0-387-15679-8 Springer-Verlag New York Heidelberg Berlin Tokyo

Library of Congress Cataloging-in-Publication Data. Krieg, Aloys, 1955 – Modular forms on half-spaces of quaternions. (Lecture notes in mathematics; 1143) Bibliography: p. Includes index. 1. Forms, Modular. 2. Quaternions. I. Title II. Series: Lecture notes in mathematics (Springer-Verlag; 1143. QA.L28 no. 1143 520 s 85-17284 [QA243] [512'.522]
ISBN 0-387-15679-8 (U.S.)

Printing and binding: Beltz Offsetdruck, Hemsbach/Bergstr.
2146/3140-543210

Table of contents

Introduction		V
Notations		XIII
Chapter I	Integral and Hermitian matrices	1
§1	Orderings	2
§2	Integral matrices	14
§3	Hermitian matrices	20
§4	MINKOWSKI's reduction theory	28
§5	Applications of reduction theory	37
Chapter II	Modular group and fundamental domain	42
§1	Symplectic group and half-space	43
§2	Modular group	54
§3	The fundamental domain $F(n;\mathbb{F})$	58
§4	Congruence subgroups	67
Chapter III	Modular forms	71
§1	Analytic class invariants	72
§2	The vector space of modular forms	83
§3	Modular forms with respect to congruence subgroups	92
§4	Relations between SIEGEL modular forms, Hermitian modular forms and modular forms of quaternions	96
Chapter IV	Theta-series	99
§1	Elementary properties of theta-series	100
§2	Theta-series as modular forms	110
§3	Theta-series with respect to even quadratic forms	117
§4	Singular modular forms	126
Chapter V	EISENSTEIN- and POINCARÉ -series	136
§1	Integrals	137
§2	EISENSTEIN-series	145
§3	POINCARÉ -series	157
§4	Metrization	166
§5	Algebraical independence	174

Chapter VI	Modular functions	182
§1	The field of modular functions	183
§2	Modular functions with respect to congruence subgroups and symmetric modular functions	189
§3	Relations between SIEGEL modular functions, Hermitian modular functions and modular functions of quaternions	192
Bibliography		196
List of symbols		200
Index		202

Introduction

The mathematical background

The theory of elliptic modular forms and functions developed from the investigation of elliptic functions in the 19th century. One proceeds from the upper half

(1) $\mathcal{H} := \{z = x + iy \in \mathbb{C} \; ; \; y > 0\}$

in $\mathbb{C}$. By means of the group

(2) $SL(2;\mathbb{R}) := \{M \in Mat(2;\mathbb{R}) \; ; \; \det M = 1\}$

one can describe the biholomorphic maps of $\mathcal{H}$ onto itself. All these maps are of course given by the transformations

(3) $z \longmapsto M\langle z\rangle := \frac{az+b}{cz+d}$, whenever $M = \begin{pmatrix} a & b \\ c & d \end{pmatrix} \in SL(2;\mathbb{R})$.

The so-called modular group

(4) $\Gamma := SL(2;\mathbb{Z})$

forms a discrete subgroup of $SL(2;\mathbb{R})$ and plays an essential part.

By a fundamental domain one means a closed set of representatives of $\mathcal{H}$ with respect to the equivalence relation induced by Γ, where every two different interior points are not equivalent. Such a domain can easily be described by

Theorem A. $\mathcal{F} := \{z = x + iy \in \mathcal{H} \; ; \; |x| \leq \frac{1}{2} \; , \; |z| \geq 1\}$ *is a fundamental domain of* $\mathcal{H}$ *with respect to the action of* Γ .

Given $k \in \mathbb{Z}$ a holomorphic function $f : \mathcal{H} \longrightarrow \mathbb{C}$ is called an *elliptic modular form of weight* k if the following two properties are satisfied:

(5) $$\begin{cases} f(M\langle z\rangle)\,(cz+d)^{-k} = f(z) & \text{for all } z \in \mathcal{H} \text{ and } M = \begin{pmatrix} a & b \\ c & d \end{pmatrix} \in \Gamma . \\ f \text{ is bounded in every domain } y \geq \rho > 0 . \end{cases}$$

By virtue of the boundedness, a FOURIER-expansion at ∞ in the form

(6) $$f(z) = \sum_{t=0}^{\infty} \alpha(t)\, e^{2\pi i t z} \; , \; z \in \mathcal{H} ,$$

is obtained. The set $[\Gamma,k]$ of all elliptic modular forms of weight k becomes a $\mathbb{C}$-vector space.

Examples are given by the EISENSTEIN-series

$$(7) \qquad E_k(z) = \sum_{\substack{(c,d)\in\mathbb{Z}\times\mathbb{Z} \\ (c,d)\neq(0,0)}} (cz+d)^{-k}, \quad z \in H,$$

which are non-identically vanishing modular forms of weight k, whenever $k > 2$ is even.

The theta-series form another class of examples.

$$(8) \qquad \Theta(z,S) := \sum_{\mathfrak{g}\in\mathbb{Z}^m} e^{\pi i \mathfrak{g}' S \mathfrak{g} z}, \quad z \in H,$$

becomes a modular form of weight $\frac{1}{2}m$, whenever $S = S^{(m)}$ is even, unimodular and positive definite.

By means of the EISENSTEIN-series, the vector space of modular forms can be described in the following well-known way:

<u>Theorem B.</u> a) <u>One has</u> $[\Gamma,k] = \{0\}$, <u>whenever</u> k <u>is negative or odd</u>. <u>If</u> $k \geq 0$ <u>is even then</u>

$$\dim\,[\Gamma,k] = \begin{cases} \left[\frac{k}{12}\right], & \text{if } k \equiv 2 \bmod 12, \\ \left[\frac{k}{12}\right] + 1, & \text{else}, \end{cases}$$

<u>holds</u>.

b) <u>Let</u> $k \geq 0$ <u>be even, then the products of</u> EISENSTEIN-<u>series</u> $E_4^l E_6^m$, <u>where</u> $l,m \in \mathbb{N}_0$ <u>and</u> $4l + 6m = k$, <u>form a basis of</u> $[\Gamma,k]$.

A meromorphic function f on H, which does not possess an essential singularity at ∞, is called an <u>elliptic modular function</u> if

$$(9) \qquad f(M\langle z\rangle) = f(z) \quad \text{for all } z \in H \text{ and } M \in \Gamma.$$

Let M denote the field of elliptic modular functions. Then

$$(10) \qquad j := \frac{1728(60E_2)^3}{(60E_2)^3 - (140E_3)^2}$$

belongs to M, and one can derive the fundamental

<u>Theorem C.</u> $\mathcal{M} = \mathbb{C}(j)$.

Especially the transcendence degree of $\mathcal{M}$ over $\mathbb{C}$ equals 1 and every modular function can be represented as a quotient of two modular forms of the same weight.

The theory of elliptic modular forms has been generalized in several respects. Replacing $\mathcal{H}$ by a product of upper half-planes one attains to the theory of HILBERT's or HILBERT-BLUMENTHAL's modular functions.

In the 30's SIEGEL [53,54] investigated the so-called modular functions of degree n , which nowadays are known as SIEGEL's modular functions. In the case n = 1 they coincide with the elliptic modular functions.

With regard to this generalization one has to replace $\mathcal{H}$ by SIEGEL's half-space

(1') $$H(n;\mathbb{R}) := \{Z = X + iY \in \mathrm{Mat}(n;\mathbb{C}) \; ; \; X=X' , Y=Y' \text{ positive definite}\}.$$

Considering the symplectic group

(2') $$Sp(n;\mathbb{R}) := \{M \in \mathrm{Mat}(2n;\mathbb{R}) \; ; \; M'JM = J\} \quad , \; J = \begin{pmatrix} 0 & I \\ -I & 0 \end{pmatrix} ,$$

instead of $SL(2;\mathbb{R})$, all biholomorphic maps of SIEGEL's half-space onto itself are given by the symplectic transformations

(3') $$Z \longmapsto M\langle Z\rangle := (AZ + B)(CZ + D)^{-1} \quad , \text{ where } \; M = \begin{pmatrix} A & B \\ C & D \end{pmatrix} \in Sp(n;\mathbb{R}) ,$$

in view of [54]. The so-called modular group

(4') $$\Gamma_n := Sp(n;\mathbb{Z})$$

forms a discrete subgroup of $Sp(n;\mathbb{R})$, which acts discontinuously on SIEGEL's half-space. By means of MINKOWSKI's reduction theory SIEGEL [54] constructed a fundamental domain:

<u>Theorem A'.</u>

$$F_n := \left\{ Z = X + iY \in H(n;\mathbb{R}) \; ; \; \begin{array}{l} X \bmod 1 \; , \; Y \; \underline{\text{reduced}} \; , \\ |\det(CZ+D)| \geq 1 \; \underline{\text{for}} \; M \in \Gamma_n \end{array} \right\}$$

<u>is a fundamental domain of</u> $H(n;\mathbb{R})$ <u>with respect to the action of the modular group</u> Γ_n .

Let $k \in \mathbb{Z}$, $n \geq 2$ and $f : H(n;\mathbb{R}) \longrightarrow \mathbb{C}$ be holomorphic. f is called SIEGEL's <u>modular form of weight</u> k if

(5') $f(M\langle Z\rangle)\det(CZ+D)^{-k} = f(Z)$

holds for all $Z \in H(n;\mathbb{R})$ and $M = \begin{pmatrix} A & B \\ C & D \end{pmatrix} \in \Gamma_n$. KOECHER [35] discovered that any requirement of boundedness is not necessary, whenever $n \geq 2$. A modular form f possesses a FOURIER-expansion of the form

(6') $f(Z) := \sum_T \alpha(T)\, e^{2\pi i \operatorname{trace} TZ}$, $Z \in H(n;\mathbb{R})$,

where T runs through the set of all positive semi-definite, semi-integral $n \times n$ matrices.

SIEGEL [54] showed that the vector space $[\Gamma_n,k]$ of all modular forms of weight k has finite dimension and consists only of 0 , if k is negative. But even nowadays it seems to be impossible to compute the exact dimension or, moreover, to state a basis for arbitrarily given k and n . The case $n = 2$ however was solved on the analogy of Theorem B by IGUSA [26] and later by FREITAG [17].

MAASS [43;44] generalized PETERSSON's scalar product to SIEGEL'S modular forms. By virtue of this theory of metrization he succeeded in proving a theorem of representation, which says that $[\Gamma_n,k]$ is generated by POINCARÉ -series, whenever $k > 2n$ is even. Another version of the theorem of representation by means of generalizations of EISENSTEIN-series (7) is due to KLINGEN [32].

Theta-series on SIEGEL's half-space are built in the form

(8') $\Theta(Z,S) := \sum_{G \in \mathrm{Mat}(m,n;\mathbb{Z})} e^{\pi i \operatorname{trace}(G'SGZ)}$, $Z \in H(n;\mathbb{R})$.

If $S = S^{(m)}$ is even, unimodular and positive definite $\Theta(\cdot,S)$ belongs to $[\Gamma_n,\frac{1}{2}m]$. RESNIKOFF [48,49] and FREITAG [19,20] showed that the theta-series form a basis of $[\Gamma_n,k]$, whenever $0 < k < \frac{1}{2}n$. The meaning of theta-series is emphasized by BÖCHERER's result [5] that the theta-series span the vector space $[\Gamma_n,k]$, whenever $k > 2n$ and $k \equiv 0 \bmod 4$.

Summarizing one has

<u>Theorem B'.</u> a) <u>Every</u> SIEGEL'<u>s</u> <u>modular form of weight</u> $k < 0$ <u>vanishes identically and</u> $\dim[\Gamma_n,k] < \infty$ <u>holds for</u> $k \geq 0$.

b) <u>If</u> $0 < k < \frac{1}{2}n$ <u>the theta-series form a basis and for</u> $k > 2n$, $k \equiv 0 \bmod 4$ <u>they span</u> $[\Gamma_n,k]$.

c) POINCARÉ - <u>and</u> EISENSTEIN-<u>series span</u> $[\Gamma_n,k]$, <u>whenever</u> $k > 2n$ <u>is even</u>.

If $n \geq 2$ a meromorphic function f on $H(n;\mathbb{R})$ satisfying

$$(9') \qquad f(M\langle Z\rangle) = f(Z) \quad \text{for all} \quad Z \in H(n;\mathbb{R}) \ , \ M \in \Gamma_n$$

is called SIEGEL's <u>modular function</u>. The set M_n of all modular functions on $H(n;\mathbb{R})$ forms a field of transcendence degree $\frac{1}{2}n(n+1)$. First it was shown by means of deep-rooted results of the algebraic geometry and later on by SIEGEL [57] using classic methods that every modular function can be represented as the quotient of two modular forms of the same weight. Finally one has the fundamental

<u>Theorem C'.</u> <u>There are SIEGEL's modular functions</u> $f_0,\ldots,f_h$, $h = \frac{1}{2}n(n+1)$, <u>satisfying</u>

$$M_n = \mathbb{C}(f_0,\ldots,f_h) \ .$$

Hermitian modular functions, which were introduced by BRAUN [7], represent another generalization of elliptic modular functions. Results by analogy with SIEGEL's theory are due to BRAUN [7,8] and KLINGEN [28,29].

All the quoted theories are special cases of a theory of automorphic forms and functions with respect to a discontinuously acting group with non-compact fundamental domain.

<u>The contents of this volume</u>

In this volume the theory of modular forms and functions on the half-space of quaternions is developed as another generalization of the classical theory of elliptic modular forms and functions. Under the objective of attaining to explicit results, it is mainly "SIEGEL's methods", which are used to derive analogous assertions to the case of SIEGEL's modular forms.

Let $\mathbb{H}$ denote the skew field of real quaternions and "-" its canonical involution. Thus the half-space of quaternions

$$(1'') \qquad H(n;\mathbb{H}) := \left\{ Z = X + iY \in \mathrm{Mat}(n;\mathbb{H}) \otimes_{\mathbb{R}} \mathbb{C} \ ; \ \begin{array}{l} X=\overline{X}' \ , \ Y=\overline{Y}' \ , \\ Y \ \text{positive definite} \end{array} \right\}$$

is defined as a subset of the tensor product $\mathrm{Mat}(n;\mathbb{H}) \otimes_{\mathbb{R}} \mathbb{C}$. In the case $n = 1$ one obtains the upper half in $\mathbb{C}$:

$$H(1;\mathbb{H}) = H(1;\mathbb{R}) = H \ .$$

On the one hand SIEGEL's half-space $H(n;\mathbb{R})$ and the Hermitian half-space $H(n;\mathbb{C})$ can be considered as analytic submanifolds of $H(n;\mathbb{H})$. On the other hand it sometimes proves useful to imbed $H(n;\mathbb{H})$ into $H(2n;\mathbb{C})$ or $H(4n;\mathbb{R})$.

The symplectic group (2') is generalized to

(2") $\quad Sp(n;\mathbb{H}) = \{M \in Mat(2n;\mathbb{H})\ ;\ \bar{M}'JM = J\}$.

Hence all biholomorphic maps of $H(n;\mathbb{H})$ onto itself are given by symplectic transformations

(3") $\quad Z \longmapsto M\langle Z\rangle := (AZ+B)(CZ+D)^{-1}, \quad M = \begin{pmatrix} A & B \\ C & D \end{pmatrix} \in Sp(n;\mathbb{H})$,

where in the case $n = 2$ one has to add the transposed map $Z \longmapsto Z'$.

The modular group is defined by

(4") $\quad \Gamma_n := Sp(n;\mathcal{O})$,

where $\mathcal{O}$ denotes the quaternions of HURWITZ [25]. Γ_n forms a discrete subgroup of $Sp(n;\mathbb{H})$ and acts discontinuously on $H(n;\mathbb{H})$. The validity of the Euclidean algorithm in $\mathcal{O}$ turns out to be the decisive factor to adopt results from SIEGEL's theory. The difficulties arising from the number theory, if one considers Hermitian modular forms, do not occur.

Now one defines $\det(CZ+D)^k$ for $Z \in H(n;\mathbb{H})$, $M = \begin{pmatrix} A & B \\ C & D \end{pmatrix} \in Sp(n;\mathbb{H})$ and even $k \in \mathbb{Z}$ by means of the embedding of $H(n;\mathbb{H})$ into $H(2n;\mathbb{C})$. In virtue of a generalization of MINKOWSKI's reduction theory, one can proceed as SIEGEL [54] did in order to construct a fundamental domain.

Theorem A". $\quad F_n := \left\{ Z = X + iY \in H(n;\mathbb{H})\ ;\ \begin{array}{l} X \bmod 1,\ Y \text{ reduced} \\ |\det(CZ+D)|^2 \geq 1 \text{ for } M \in \Gamma_n \end{array} \right\}$

is a fundamental domain of $H(n;\mathbb{H})$ *with respect to the action of the modular group* Γ_n.

Given $n \geq 2$ and an even $k \in \mathbb{Z}$ a *modular form of quaternions of weight* k is defined by analogy to (5') and possesses a FOURIER-expansion corresponding to (6').

Classic procedure yields that the vector space $[\Gamma_n,k]$ of modular forms of quaternions having weight k proves finitely dimensional and consists only of 0 , whenever $k < 0$.

By applying PETERSSON's resp. MAASS' theory of metrization to modular forms of quaternions, theorems of representation are obtained. $[\Gamma_n,k]$ is spanned by generalized EISENSTEIN- and POINCARÉ -series, whenever $k > 8n-6$ is even.

Theta-series are defined in correspondence to (8') and the theory of singular modular forms yields a basis of $[\Gamma_n,k]$ consisting of theta-series under the condition $0 < k < 2n$.

Summarizing one has

Theorem B". a) Every modular form of quaternions having negative weight vanishes identically and

$$\dim\,[\Gamma_n,k] < \infty$$

holds, whenever $k \geq 0$ is even.

b) If $0 < k < 2n$ theta-series form a basis of $[\Gamma_n,k]$.

c) Let $k > 8n-6$ be even, then $[\Gamma_n,k]$ is spanned by EISENSTEIN- and POINCARÉ -series.

If $n \geq 2$ modular functions of quaternions are defined on the analogy of (9') to be meromorphic functions on $H(n;\mathbb{H})$, which remain invariant under all modular transformations. The set M_n of all modular functions of quaternions forms a field of transcendence degree $n(2n-1)$. Modular functions again can be represented as quotients of modular forms having the same weight and one achieves

Theorem C". There exist modular functions of quaternions $f_o,\dots,f_h$, $h = n(2n-1)$, such that

$$M_n = \mathbb{C}(f_o,\dots,f_h) \ .$$

In view of the choice of the quaternions of HURWITZ as integral elements, it is possible to represent the theory of modular forms of quaternions, of Hermitian modular forms with respect to the Gaussian integers and of SIEGEL's modular forms throughout the whole volume simultaneously.

The method of compactification according to SATAKE [51] resp. BAILY and BOREL [4] will not be treated in the sequel. The theory of HECKE-operators is also left out of consideration, since the situation is completely different from that dealing with SIEGEL's modular forms. But an investigation of this subject will follow [41].

My thanks are due to Professor Dr. M. KOECHER for his encouragement and helpful advice.

Notations

Let $\mathbb{N}$ denote the natural numbers without 0 , $\mathbb{Z}$ the integral, $\mathbb{Q}$ the rational, $\mathbb{R}$ the real, $\mathbb{C}$ the complex numbers and $\mathbb{H}$ the skew field of real quaternions. Given a ring R and $m,n \in \mathbb{N}$ let $Mat(m,n;R)$ denote the set of matrices having m rows and n columns and coefficients in R . We use the abbreviations $Mat(n;R) := Mat(n,n;R)$ and $R^m := Mat(m,1;R)$. On the other hand $A^{(m,n)}$ also stands for a matrix having m rows and n columns as well as $A^{(n)}$ for $A^{(n,n)}$. If $A_1,\ldots,A_n$ are quadratic matrices $[A_1,\ldots,A_n]$ is defined to be that quadratic matrix, where $A_1,\ldots,A_n$ stand on the diagonal and 0's else. A' always denotes the transposed matrix of A . The letter $I = I^{(n)}$ is reserved for the identity matrix $[1,\ldots,1]$, whereas 0 stands for a matrix of suitable size having only 0's . The identical map of a non-empty set M onto itself is denoted by id_M . By Υ_n we mean the symmetric group on n letters, i.e. the group of all bijective maps of $\{1,\ldots,n\}$ onto itself. The KRONECKER-symbol is denoted by δ_{jk} and $ord\, M$ equals the number of elements of a set M.

Finally the way of citing is explained. The whole volume is divided into 6 chapters and each chapter into paragraphs. Throughout each paragraph the theorems, lemmata and propositions are numbered consecutively. Only if assertions of another chapter are quoted the chapter is cited by Roman numerals. The numbers in brackets refer to the bibliography at the end of this volume.

Chapter I Integral and Hermitian matrices

In this chapter the notion of an integral element is introduced for the fields $\mathbb{R}$ and $\mathbb{C}$ as well as the skew field $\mathbb{H}$ of real quaternions. This is done by the choice of a special ordering. Whereas one surveys the arithmetical properties of these orderings over $\mathbb{R}$ and $\mathbb{C}$ because one deals with commutative Euclidean rings, the number theory of the quaternions of HURWITZ has to be pointed out in greater detail.

By means of this ordering, integral and unimodular matrices are defined. An introduction to the arithmetical theory of integral matrices is given in §2. Special attention is devoted to a weak version of the elementary divisor theorem over the non-commutative ring of the quaternions of HURWITZ.

§3 deals with the analysis of Hermitian matrices. Especially, the properties of positive definite matrices are examined.

The reduction theory of MINKOWSKI is generalized in §4. A fundamental domain of the space of positive definite Hermitian matrices with respect to the action of the unimodular group is determined.

Finally §5 presents some useful applications of the "Main theorem of the reduction theory". Especially, the class numbers are estimated. Furthermore the groups of automorphs of positive definite matrices are examined.

§1 Orderings

Throughout the volume let $\mathbb{F}$ denote the field $\mathbb{R}$ of real numbers or the field $\mathbb{C}$ of complex numbers or the skew field $\mathbb{H}$ of real quaternions [27], 2.4. The letter $r = r(\mathbb{F})$ always stands for the dimension of $\mathbb{F}$ regarded as a vector space over $\mathbb{R}$, hence

$$r(\mathbb{R}) = 1 \ , \ r(\mathbb{C}) = 2 \ , \ r(\mathbb{H}) = 4 \ .$$

We consider the inclusions

$$\mathbb{R} \subset \mathbb{C} \subset \mathbb{H}$$

and denote the canonical basis of $\mathbb{F}$ over $\mathbb{R}$ by $e_1 = 1, \ldots, e_r$. The product in $\mathbb{F}$ is given by $\mathbb{R}$-linear extension of the products in the following multiplication table:

$$e_1 e_j = e_j e_1 = e_j \ , \ 1 \leq j \leq r \ ,$$
$$e_j^2 = -e_1 \ , \ 2 \leq j \leq r \ ,$$

and for $\mathbb{F} = \mathbb{H}$ additionally by

$$e_2 e_3 = -e_3 e_2 = e_4 \ , \ e_3 e_4 = -e_4 e_3 = e_2 \ , \ e_4 e_2 = -e_2 e_4 = e_3 \ .$$

Each $a \in \mathbb{F}$ possesses a unique representation

(1) $$a = \sum_{j=1}^{r} a_j e_j \ , \text{ where } a_j \in \mathbb{R} \ .$$

Applying this representation we can define the following maps:

$$\mathrm{Re} : \mathbb{F} \longrightarrow \mathbb{R} \ , \ a \longmapsto a_1 \ ,$$
$$N : \mathbb{F} \longrightarrow \mathbb{R} \ , \ a \longmapsto \sum_{j=1}^{r} a_j^2 \ ,$$
$$\bar{\ } : \mathbb{F} \longrightarrow \mathbb{F} \ , \ a \longmapsto \bar{a} := 2\,\mathrm{Re}(a) - a \ .$$

$\mathrm{Re}(a)$ is called the _real part_ of a, $N(a)$ the _norm_ of a and $\bar{a}$ the _conjugate_ of a.

Elementary computations using the above multiplication table imply the following well-known properties of these maps.

Proposition 1.1. a) _The map_ $\bar{\ } : \mathbb{F} \longrightarrow \mathbb{F}$, $a \longmapsto \bar{a}$, _is an involution, i.e._ $\overline{\alpha a + \beta b} = \alpha\bar{a} + \beta\bar{b}$, $\overline{ab} = \bar{b}\bar{a}$ _and_ $\bar{\bar{a}} = a$ _for_ $a, b \in \mathbb{F}$ _and_ $\alpha, \beta \in \mathbb{R}$.

b) _Given_ $a, b \in \mathbb{F}$ _of the form_ (1) _the following holds:_

$$\mathrm{Re}(a) = \mathrm{Re}(\bar{a}) = \frac{1}{2}(a + \bar{a}) \ ,$$

$$\mathrm{Re}(\bar{a}b) = \mathrm{Re}(b\bar{a}) = \sum_{j=1}^{r} a_j b_j ,$$
$$N(a) = N(\bar{a}) = a\bar{a} ,$$
$$N(ab) = N(a)N(b) ,$$
$$a^2 - 2\,\mathrm{Re}(a)\,a + N(a) = 0 .$$

As usual one defines the <u>center</u> of $\mathbb{F}$ by

$$\mathrm{Cent}\,\mathbb{F} = \{a \in \mathbb{F} \; ; \; ab = ba \quad \text{for} \quad b \in \mathbb{F}\}$$

and achieves

$$\mathrm{Cent}\,\mathbb{R} = \mathbb{R} , \quad \mathrm{Cent}\,\mathbb{C} = \mathbb{C} , \quad \mathrm{Cent}\,\mathbb{H} = \mathbb{R}$$

in the cases under consideration.

The arithmetical properties essentially depend on the definition of integral elements in $\mathbb{F}$. We proceed by specifying an ordering.

<u>Definition.</u> Let V be a vector space over $\mathbb{R}$ of dimension $n < \infty$. A subset L of V is called a <u>lattice</u> in V if there exists a basis $v_1, \ldots, v_n$ of V such that

$$L = \mathbb{Z}v_1 + \ldots + \mathbb{Z}v_n .$$

<u>Definition.</u> A subset $\mathcal{O} = \mathcal{O}(\mathbb{F})$ of $\mathbb{F}$ is called an <u>ordering</u> of $\mathbb{F}$ if it satisfies the following properties:

(O.1) $\mathcal{O}$ is a lattice in $\mathbb{F}$.

(O.2) $\mathcal{O}$ is a subring of $\mathbb{F}$ containing 1.

(O.3) $2\,\mathrm{Re}(a) \in \mathbb{Z}$ and $N(a) \in \mathbb{Z}$ for $a \in \mathcal{O}$.

An ordering $\mathcal{O}$ is said to be <u>maximal</u> if any ordering containing $\mathcal{O}$ already equals $\mathcal{O}$.

In view of Proposition 1.1 property (O.3) means that every $a \in \mathcal{O}(\mathbb{F})$ is a root of the polynomial

$$p(X) = X^2 - 2\,\mathrm{Re}(a)\,X + N(a) \in \mathbb{Z}[X] .$$

Using (O.2) and (O.3) one has

$$\bar{a} \in \mathcal{O} \quad \text{for} \quad a \in \mathcal{O} .$$

Whereas $\mathbb{Z}$ is the only ordering in $\mathbb{R}$, there are "many" orderings in $\mathbb{C}$ and $\mathbb{H}$. The ring of integers of any imaginary-quadratic number field forms a maximal ordering of $\mathbb{C}$. For orderings of the quaternions the reader is referred to [59].

In this volume one special maximal ordering is fixed for $\mathbb{C}$ resp. $\mathbb{H}$. The arithmetic essentially depends on the choice of this ordering.

We define $\mathcal{O} = \mathcal{O}(\mathbb{F})$ by

$$\mathcal{O}(\mathbb{R}) = \mathbb{Z},$$
$$\mathcal{O}(\mathbb{C}) = \mathbb{Z}e_1 + \mathbb{Z}e_2,$$
$$\mathcal{O}(\mathbb{H}) = \mathbb{Z}e_o + \mathbb{Z}e_1 + \mathbb{Z}e_2 + \mathbb{Z}e_3, \text{ where } e_o = \tfrac{1}{2}(e_1 + e_2 + e_3 + e_4).$$

The elements of $\mathcal{O}(\mathbb{C})$ are known as GAUSS<u>ian integers</u>, the elements of $\mathcal{O}(\mathbb{H})$ as the <u>quaternions of</u> HURWITZ, since their arithmetical properties were examined by HURWITZ [25]. For our applications we need results which extend beyond those of HURWITZ. The basic properties are therefore summarized below.

In the following the elements of $\mathcal{O}(\mathbb{F})$ are called <u>integral</u>.

<u>Theorem 1.2.</u> <u>$\mathcal{O}(\mathbb{F})$ is a maximal ordering of $\mathbb{F}$.</u>

<u>Proof.</u> The assertion is obvious for $\mathbb{F} = \mathbb{R}, \mathbb{C}$. In order to prove that $\mathcal{O} = \mathcal{O}(\mathbb{H})$ is an ordering it suffices to show that $\mathcal{O}$ is closed under multiplication by e_o. Considering the multiplication table above one easily verifies:

$$e_oe_1 = e_1e_o = e_o, \qquad e_o^2 = e_o - e_1,$$
$$e_oe_2 = -e_o + e_2 + e_3, \quad e_2e_o = e_o - e_1 - e_3,$$
$$e_oe_3 = e_o - e_1 - e_2, \quad e_3e_o = -e_o + e_2 + e_3.$$

Thus $\mathcal{O}$ is an ordering of $\mathbb{H}$.

Let L denote an arbitrary ordering containing $\mathcal{O}$. Given $a \in L$ of the form (1) one has $a + e_j \in L$, hence $N(a + e_j) \in \mathbb{Z}$ for $1 \leq j \leq 4$ Now $N(a) = a_1^2 + a_2^2 + a_3^2 + a_4^2 \in \mathbb{Z}$ yields $2a_j \in \mathbb{Z}$ for $1 \leq j \leq 4$, where $2a_1, \ldots, 2a_4$ are either all even or all odd. Thus $a \in \mathcal{O}$ and $L = \mathcal{O}$. □

We denote the set of <u>units</u> of $\mathcal{O}(\mathbb{F})$ by $E = E(\mathbb{F})$. Since $\mathcal{O}$ is a unital ring, E forms a group. E consists of all elements $a \in \mathcal{O}$ for which there exists $b \in \mathcal{O}$ such that $ab = ba = 1$. For $0 \neq a \in \mathbb{F}$ one has $a^{-1} = N(a)^{-1}\bar{a}$, hence

$$E(\mathbb{F}) = \{a \in \mathcal{O}(\mathbb{F})\ ;\ N(a) = 1\}.$$

Using this description one easily verifies

$$E(\mathbb{R}) = \{\pm 1\},$$
$$E(\mathbb{C}) = \{\pm e_1, \pm e_2\},$$
$$E(\mathbb{H}) = \{\pm e_1, \pm e_2, \pm e_3, \pm e_4, \tfrac{1}{2}(\pm e_1 \pm e_2 \pm e_3 \pm e_4)\}.$$

The group of units $E(\mathbb{F})$ acts on $\mathbb{F}$ by multiplication from the left and from the right. Whereas these actions can easily be described for $\mathbb{F} = \mathbb{R},\mathbb{C}$ the situation has to be examined carefully for the quaternions.

By $S(\mathbb{F})$ we mean the set of units lying in the center of $\mathbb{F}$, i.e.

$$S(\mathbb{F}) = E(\mathbb{F}) \cap \mathrm{Cent}\,\mathbb{F} ,$$

$$S(\mathbb{R}) = \{\pm 1\} , \quad S(\mathbb{C}) = \{\pm e_1, \pm e_2\} , \quad S(\mathbb{H}) = \{\pm 1\} .$$

The letter $s := s(\mathbb{F})$ always denotes the number of elements of $S(\mathbb{F})$, hence

$$s(\mathbb{R}) = 2 , \quad s(\mathbb{C}) = 4 , \quad s(\mathbb{H}) = 2 .$$

Given an element of $S(\mathbb{F})$ the actions of multiplication from the left and from the right coincide.

$E(\mathbb{H})$ contains the quaternion group

$$\mathcal{Q} = \{\pm e_1, \pm e_2, \pm e_3, \pm e_4\}$$

as a subgroup. One easily verifies that $\mathcal{Q}$ is a normal subgroup of index 3 and that $e_1, e_o, \bar{e}_o$ form a set of representatives of the cosets.

<u>Lemma 1.3.</u> <u>The group of conjugations</u>

$$\mathbb{H} \longrightarrow \mathbb{H} , \quad a \longmapsto \bar{\rho} a \rho , \quad \rho \in E(\mathbb{H}) ,$$

<u>coincides with the group of maps</u>

$$(2) \qquad \mathbb{H} \longrightarrow \mathbb{H} , \quad a = \sum_{j=1}^{4} a_j e_j \longmapsto \sum_{j=1}^{4} \varepsilon_j a_{\pi(j)} e_j ,$$

<u>where</u> π <u>runs through the set of even permutations in</u> γ_4 <u>satisfying</u> $\pi(1) = 1$ <u>and</u> $(\varepsilon_1, \dots, \varepsilon_4)$ <u>through those quadruples in</u> $\{\pm 1\}^4$ <u>such that</u> $\varepsilon_1 = \varepsilon_2 \cdot \varepsilon_3 \cdot \varepsilon_4 = 1$.

<u>Proof.</u> Given $a \in \mathbb{H}$ of the form (1) one easily computes

$$\bar{e}_2 a e_2 = a_1 e_1 + a_2 e_2 - a_3 e_3 - a_4 e_4 ,$$

$$\bar{e}_3 a e_3 = a_1 e_1 - a_2 e_2 + a_3 e_3 - a_4 e_4 ,$$

$$\bar{e}_4 a e_4 = a_1 e_1 - a_2 e_2 - a_3 e_3 + a_4 e_4 ,$$

$$\bar{e}_o a e_o = a_1 e_1 + a_3 e_2 + a_4 e_3 + a_2 e_4 ,$$

$$e_o a \bar{e}_o = a_1 e_1 + a_4 e_2 + a_2 e_3 + a_3 e_4 .$$

The above remarks concerning the quaternion group in $E(\mathbb{H})$ yield that any conjugation can be described in the form (2). Two conjugations

$a \longmapsto \bar{\rho} a \rho$ and $a \longmapsto \bar{\delta} a \delta$ coincide if and only if $\delta\rho^{-1} \in \mathcal{S}(\mathbb{H}) = \{\pm 1\}$. Hence the group of conjugations consists of 12 elements. But there are also exactly 12 maps of the form (2). Thus both groups coincide. □

By means of the above considerations we can construct a fundamental domain of $\mathbb{F}$ with respect to the action of $E(\mathbb{F})$. Therefore we need the more general

<u>Definition.</u> Let T be a topological space and G a subgroup of $\mathrm{Aut}\, T := \{f : T \longrightarrow T \, ; \, f \text{ homeomorphism}\}$. A subset F of T is called a <u>fundamental domain of T with respect to G</u> if it fulfills the following conditions:

(F.1) F is closed and possesses interior points.

(F.2) Given $x \in T$ there exists $g \in G$ such that $g(x) \in F$.

(F.3) If x and $g(x)$ for some $g \in G$ are interior points of F it follows that $g = \mathrm{id}_T$.

(F.4) Given a compact subset C of T there exist only finitely many $g \in G$ such that $g(F) \cap C \neq \emptyset$.

(F.5) The set $\{g \in G \, ; \, g(F) \cap F \neq \emptyset\}$ is finite.

The images $g(F)$, $g \in G$, satisfying $g(F) \cap F \neq \emptyset$ are called <u>neighbors</u> of F.

Now we define

$$F(\mathbb{F}) := \{a \in \mathbb{F} \, ; \, \mathrm{Re}(a) \geq \mathrm{Re}(\bar{\varepsilon} a) \quad \text{for} \quad \varepsilon \in E(\mathbb{F})\} .$$

Using the explicit description of $E(\mathbb{F})$ and Proposition 1.1 yields

$$F(\mathbb{R}) = \{a \in \mathbb{R} \, ; \, a \geq 0\} ,$$

$$F(\mathbb{C}) = \{a = a_1 e_1 + a_2 e_2 \in \mathbb{C} \, ; \, a_1 \geq |a_2|\} ,$$

$$F(\mathbb{H}) = \{a = \sum_{j=1}^{4} a_j e_j \in \mathbb{H} \, ; \, a_1 \geq |a_2| + |a_3| + |a_4|\} .$$

Additionally we put

$$F^*(\mathbb{H}) = \{a \in F(\mathbb{H}) \, ; \, a_2 \geq a_3 \geq 0 \, , \, a_2 \geq |a_4|\} ,$$

$$F^*(\mathbb{F}) = F(\mathbb{F}) \quad \text{for} \quad \mathbb{F} = \mathbb{R}, \mathbb{C} .$$

<u>Theorem 1.4.</u> a) <u>$F(\mathbb{F})$ is a fundamental domain of $\mathbb{F}$ with respect to the action of $E(\mathbb{F})$ by multiplication from the right or from the left.</u>

b) $F^*(\mathbb{F})$ <u>is a fundamental domain of</u> $\mathbb{F}$ <u>with respect to the action of</u> $E(\mathbb{F})^2$: $\mathbb{F} \longrightarrow \mathbb{F}$, $a \longmapsto \varepsilon a \delta$, $\varepsilon, \delta \in E(\mathbb{F})$.

<u>Proof.</u> It suffices to consider the case $\mathbb{F} = \mathbb{H}$ and to prove conditions (F.2) and (F.3). Given $a \in \mathbb{H}$ choose $\varepsilon \in E(\mathbb{H})$ such that

$$\mathrm{Re}(\bar{\varepsilon}a) = \max\{\mathrm{Re}(\bar{\delta}a);\ \delta \in E(\mathbb{H})\}$$

therefore $\bar{\varepsilon}a \in F(\mathbb{H})$. According to Lemma 1.3 one can find $\delta \in E(\mathbb{H})$ such that $\bar{\delta}\bar{\varepsilon}a\delta \in F^*(\mathbb{H})$.

Now consider $\varepsilon, \delta \in E(\mathbb{H})$ such that a and $\varepsilon a \delta$ belong to the interior of $F^*(\mathbb{H})$. From $F^*(\mathbb{H}) \subset F(\mathbb{H})$ we get $\mathrm{Re}(\bar{\rho}a) > \mathrm{Re}(a)$ for $1 \neq \rho \in E(\mathbb{H})$. The same considerations for $\varepsilon a \delta$ together with $\mathrm{Re}(\varepsilon a \delta) = \mathrm{Re}(\delta \varepsilon a)$ imply $\delta\varepsilon = 1$, hence $\bar{\delta} = \varepsilon$. Since a belongs to the interior of $F^*(\mathbb{H})$ we have $a_2 > a_3 > 0$ and $a_2 > |a_4|$. Now Lemma 1.3 yields $\delta = \pm 1$.

Given $\varepsilon \in E(\mathbb{H})$ such that a and εa belong to the interior of $F(\mathbb{H})$, one proves $\varepsilon = 1$ in an analogous way. □

Following COHN [14],3.1, we call $0 \neq a \in \mathcal{O}$ an <u>invariant element</u> if $a\mathcal{O} = \mathcal{O}a$, i.e. if the map

$$\mathcal{O} \longrightarrow \mathcal{O},\quad x \longmapsto axa^{-1},$$

is bijective. Put

$$I(\mathcal{O}) := \{a \in \mathcal{O};\ a \text{ invariant}\}.$$

Clearly

$$I(\mathcal{O}(\mathbb{F})) = \mathcal{O}(\mathbb{F}) - \{0\} \quad \text{for } \mathbb{F} = \mathbb{R}, \mathbb{C}.$$

<u>Lemma 1.5.</u> $I(\mathcal{O}(\mathbb{H})) = \{\alpha\varepsilon, \alpha(e_1+e_2)\varepsilon;\ \alpha \in \mathbb{N},\ \varepsilon \in E\}$.

<u>Proof.</u> Obviously the elements $\alpha\varepsilon$, $\alpha \in \mathbb{N}$, $\varepsilon \in E$, belong to $I(\mathcal{O})$. In order to prove that the elements quoted above are invariant it suffices to demonstrate this property for $e_1 + e_2$, because $I(\mathcal{O})$ is a semigroup. One easily calculates:

$$(e_1 + e_2)e_1(e_1 + e_2)^{-1} = e_1,\quad (e_1 + e_2)e_2(e_1 + e_2)^{-1} = e_2,$$
$$(e_1 + e_2)e_3(e_1 + e_2)^{-1} = e_4,\quad (e_1 + e_2)e_4(e_1 + e_2)^{-1} = -e_3.$$

Hence $e_1 + e_2$ is invariant, too.

For any $x \in \mathbb{H}$ it follows that

$$2(\mathrm{Re}(x))^2 - N(x) = \tfrac{1}{2}(x + \bar{x})^2 - x\bar{x} = \tfrac{1}{2}(x^2 + \bar{x}^2) = \mathrm{Re}(x^2).$$

Therefore we have $x^2 = -1$ if and only if $\mathrm{Re}(x) = 0$ and $N(x) = 1$. Hence $g^2 = -1$ holds for $g \in \mathcal{O}$ if and only if $g = \pm e_j$, $j = 2,3,4$.

Given an invariant element $0 \neq a \in \mathcal{O}$ the map

$$\varphi : \mathcal{O} \longrightarrow \mathcal{O} \,, \quad x \longmapsto axa^{-1} \,,$$

is bijective and satisfies $(\varphi(e_j))^2 = -1$ for $j = 2,3,4$. Hence there exist $\pi \in \Upsilon_4$ and $(\varepsilon_1,\ldots,\varepsilon_4) \in \{\pm 1\}^4$ such that $\pi(1) = 1$, $\varepsilon_1 = 1$ and

$$\varphi(e_j) = \varepsilon_j \, e_{\pi(j)} \,, \quad 1 \leq j \leq 4 \,.$$

From $e_2 e_3 e_4 = -1$ it follows that

$$-1 = \varphi(e_2)\varphi(e_3)\varphi(e_4) = \varepsilon_2\varepsilon_3\varepsilon_4 \, e_{\pi(2)}e_{\pi(3)}e_{\pi(4)} = -\,\mathrm{sgn}(\pi)\,\varepsilon_2\varepsilon_3\varepsilon_4 .$$

Therefore φ belongs to the group of maps

$$(3) \qquad \mathcal{O} \longrightarrow \mathcal{O} \,, \quad x = \sum_{j=1}^{4} x_j e_j \quad \longmapsto \quad \sum_{j=1}^{4} \varepsilon_j x_j e_{\pi(j)} \,,$$

where $\pi \in \Upsilon_4$, $(\varepsilon_1,\ldots,\varepsilon_4) \in \{\pm 1\}^4$ satisfying $\pi(1) = 1$, $\varepsilon_1 = 1$ and $\varepsilon_2\varepsilon_3\varepsilon_4 = \mathrm{sgn}(\pi)$. This group consists of 24 elements.

The above calculations and Lemma 1.3 show that each map of the form (3) can be described by $x \longmapsto axa^{-1}$, where $a \in E$ or $a \in (e_1 + e_2)E$. An invariant element a induces the identity map if and only if $a \in \mathbb{Z} - \{0\}$. Hence the assertion follows. □

Next we want to prove that the Euclidean algorithm also applies to the quaternions of HURWITZ.

<u>Proposition 1.6.</u> <u>Given</u> $a \in \mathbb{H}$ <u>there exists</u> $g \in \mathcal{O}(\mathbb{H})$ <u>such that</u>

$$a - g = \sum_{j=1}^{4} b_j e_j \,, \quad |b_j| \leq \tfrac{1}{2} \,, \quad 1 \leq j \leq 4 \,, \quad \sum_{j=1}^{4} |b_j| \leq 1 \,.$$

<u>Proof.</u> Given a in the form (1) we may assume $|a_j| \leq \frac{1}{2}$ for $1 \leq j \leq 4$ without restriction. If $\sum_{j=1}^{4} |a_j| > 1$ one can choose $g_j \in \{\pm\frac{1}{2}\}$ such that $|a_j - g_j| = \frac{1}{2} - |a_j| \leq \frac{1}{2}$ for $1 \leq j \leq 4$. Defining $g := \sum_{j=1}^{4} g_j e_j \in \mathcal{O}$ yields the assertion. □

<u>Corollary 1.7.</u> <u>Given</u> $a \in \mathbb{F}$ <u>there exists</u> $g \in \mathcal{O}(\mathbb{F})$ <u>such that</u>

$$N(a-g) \leq c(\mathbb{F}) \,,$$

<u>where</u> $c(\mathbb{R}) = \frac{1}{4}$ <u>and</u> $c(\mathbb{C}) = c(\mathbb{H}) = \frac{1}{2}$.

<u>Proof.</u> The assertion is obvious for $\mathbb{F} = \mathbb{R},\mathbb{C}$. In view of Proposition 1.6 for $\mathbb{F} = \mathbb{H}$ it remains to show that $\varphi(b_1,b_2,b_3,b_4) := \sum_{j=1}^{4} b_j^2 \leq \frac{1}{2}$ holds for $b_j \in \mathbb{R}$, $0 \leq b_j \leq \frac{1}{2}$ and $\sum_{j=1}^{4} b_j \leq 1$. We choose $(b_1,\ldots,b_4)$ such that φ is maximal. After suitable renumbering we may suppose that $b_1 \geq b_2 \geq b_3 \geq b_4 \geq 0$. If b_3 were positive there would exist $\varepsilon > 0$ such that $b_2 + \varepsilon \leq \frac{1}{2}$ and $b_3 - \varepsilon \geq 0$. Now

$$\varphi(b_1,b_2+\varepsilon,b_3-\varepsilon,b_4) - \varphi(b_1,b_2,b_3,b_4) = 2\varepsilon(b_2 - b_3) + 2\varepsilon^2 > 0$$

yields a contradiction, hence $b_3 = 0$. Therefore $\varphi(b_1,b_2,0,0) = b_1^2 + b_2^2 \leq \frac{1}{2}$ for $0 \leq b_1,b_2 \leq \frac{1}{2}$. □

<u>Corollary 1.8.</u> <u>Euclidean algorithm</u>
<u>Given</u> $a,b \in \mathcal{O}$, $b \neq 0$, <u>there exist</u> $x,y,z,w \in \mathcal{O}$ <u>such that</u>

$$a = bx + y \quad \text{and} \quad N(y) < N(b) ,$$
$$a = zb + w \quad \text{and} \quad N(w) < N(b) .$$

<u>Proof.</u> Consider $b^{-1}a \in \mathbb{F}$ and choose $x \in \mathcal{O}$ according to the above corollary such that $N(b^{-1}a - x) < 1$. Applying Proposition 1.1 one has $N(a-bx) = N(b)N(b^{-1}a - x) < N(b)$. For the second part consider ab^{-1} . □

The validity of the Euclidean algorithm enables us to derive a result on the generators of ideals.

<u>Definition.</u> A subgroup I of a unital ring R is called a <u>left ideal</u> resp. a <u>right ideal</u> if every $a \in I$ and $x \in R$ satisfy $xa \in I$ resp. $ax \in I$. If I is a left as well as a right ideal I is said to be a <u>two-sided ideal</u>. A left resp. right resp. two-sided ideal is called <u>principal</u> if it is generated by a single element.

A subset I of $\mathcal{O}$ is a left ideal if and only if $\bar{I} := \{\bar{a} ; a \in I\}$ is a right ideal.

<u>Proposition 1.9.</u> <u>Every left resp. right ideal</u> I <u>of</u> $\mathcal{O}$ <u>is principal</u>. <u>Each two-sided ideal of</u> $\mathcal{O}$ <u>is generated by an invariant element or</u> 0 .

<u>Proof.</u> Let $I \neq \{0\}$ be a left ideal. Because of $N(g) \in \mathbb{N}$ for $0 \neq g \in \mathcal{O}$ there exists $a \in I - \{0\}$ such that $N(a)$ is minimal. Given $b \in I$ we can choose $g,h \in \mathcal{O}$ such that $b = ga + h$ and

$N(h) < N(a)$ according to Corollary 1.8. Since I is a left ideal, one has $h = b - ga \in I$, hence $h = 0$ because of the minimality of $N(a)$. Therefore $I = \mathcal{O}a$. Given a right ideal I consider $\bar{I}$. If I is a two-sided ideal, we choose a as above. The same arguments for left and right ideals yield $\mathcal{O}a = a\mathcal{O}$. Hence $0 \neq a$ is an invariant element. □

Since the divisibility theory in the quaternions of HURWITZ and in an arbitrary commutative Euclidean ring differ, we state the basic facts according to [25] and [14].

Definition. Given $a, b \in \mathcal{O}$ we call a a left divisor resp. a right divisor of b if there is $g \in \mathcal{O}$ such that $ag = b$ resp. $ga = b$. We write $a|_l b$ resp. $a|_r b$. a and b are said to be left associated resp. right associated if there is $\varepsilon \in E$ such that $a = \varepsilon b$ resp. $a = b\varepsilon$. We use the abbreviation $a \sim_l b$ resp. $a \sim_r b$.
Given $a_1, \ldots, a_n$ in $\mathcal{O}$ we call $a \in \mathcal{O}$ a greatest common right divisor of $a_1, \ldots, a_n$ ($\mathrm{gcrd}(a_1, \ldots, a_n)$), if a is a common right divisor of $a_1, \ldots, a_n$ and any common right divisor of $a_1, \ldots, a_n$ is a right divisor of a. Especially $a_1, \ldots, a_n$ are said to be relatively right-prime if 1 is a $\mathrm{gcrd}(a_1, \ldots, a_n)$. The greatest common left divisor (gcld) is defined in an analogous way.

The elementary divisibility properties are stated only for right divisors, although analogous assertions for left divisors are valid.

Proposition 1.10. The following properties hold:

(i) $a|_r b \Leftrightarrow \bar{a}|_l \bar{b}$, $a \sim_r b \Leftrightarrow \bar{a} \sim_l \bar{b}$,

(ii) $a|_r a$, $a|_r 0$ and $1|_r a$ for $a \in \mathcal{O}$,

(iii) $a|_r 1 \Leftrightarrow a \sim_l 1 \Leftrightarrow a \in E$,

(iv) $a|_r b \Rightarrow ac|_r bc$ for $c \in \mathcal{O}$,

(v) $a|_r b_j$, $1 \leq j \leq n \Rightarrow a|_r \sum_{j=1}^{n} c_j b_j$ for $c_1, \ldots, c_n \in \mathcal{O}$,

(vi) $a|_r b$ and $b|_r c \Rightarrow a|_r c$,

(vii) $a|_r b \Leftrightarrow b \in \mathcal{O}a \Leftrightarrow \mathcal{O}b \subset \mathcal{O}a$,

(viii) $a|_r b$ and $b|_r a \Leftrightarrow \mathcal{O}a = \mathcal{O}b \Leftrightarrow a \sim_l b$.

A result concerning gcrd's is stated in the following

<u>Lemma 1.11.</u> <u>Let</u> $a_1,\dots,a_n$ <u>be in</u> $\mathcal{O}$ <u>not all</u> 0 . <u>Then there exists a greatest common right divisor</u> d <u>of</u> $a_1,\dots,a_n$, <u>which can be represented in the form</u>

$$(4)\qquad d = \sum_{j=1}^{n} b_j a_j \text{ , where } b_j \in \mathcal{O} \text{ , } 1 \leq j \leq n \text{ .}$$

<u>Any</u> $\mathrm{gcrd}(a_1,\dots,a_n)$ <u>is left-associated to</u> d .

<u>Proof.</u> Let I denote the left ideal generated by $a_1,\dots,a_n$, hence $I = \mathcal{O}a_1 + \dots + \mathcal{O}a_n$. In view of Proposition 1.9 there exists $d \in \mathcal{O}$ such that $I = \mathcal{O}d$. Thus d is a common right divisor of a_j , $1 \leq j \leq n$, and possesses a representation of the form (4). Let a denote another common right divisor of $a_1,\dots,a_n$. In view of (4) and Proposition 1.10 (v) a is a right divisor of d . Hence d is a $\mathrm{gcrd}(a_1,\dots,a_n)$. If a is another $\mathrm{gcrd}(a_1,\dots,a_n)$ one has $a|_r d$ and $d|_r a$. It follows that $a \sim_l d$ from Proposition 1.10 (viii).

□

The notions of a left and right divisor do not coincide in the quaternions of HURWITZ. For example $a = e_1 + e_2 + e_3$ is a left divisor of $b = (e_1 + e_2 + e_3)(e_1 + e_2) = 2e_2 + e_3 - e_4$. But a fails to be a right divisor of b because of $ba^{-1} = \frac{1}{3}(3e_1 + e_2 + 2e_3 - 2e_4) \notin \mathcal{O}(\mathbb{H})$.

The situation changes if we consider invariant elements.

<u>Proposition 1.12.</u> <u>Let</u> $c \in \mathcal{O}$ <u>be invariant and</u> $a \in \mathcal{O}$ <u>arbitrary</u>:

a) $\qquad c|_l a \Leftrightarrow c|_r a$,

b) $\qquad a|_l c \Leftrightarrow a|_r c$.

<u>Proof.</u> a) The assertion follows from the definition of invariant elements.
b) We use the description of $I(\mathcal{O}(\mathbb{H}))$ in Lemma 1.5. From $ab = \alpha \in \mathbb{Z}$ it follows that $ba = \alpha$, if we regard $\mathrm{Re}(ab) = \mathrm{Re}(ba)$ and $N(ab) = N(ba)$. Now $ab = \alpha\varepsilon$, $\varepsilon \in E$, yields $\varepsilon b\bar{\varepsilon} a = \alpha\varepsilon$ and $ab = \alpha(e_1 + e_2)\varepsilon$ implies $(e_1 + e_2)\varepsilon b\bar{\varepsilon}(e_1 + e_2)^{-1} a = \alpha(e_1 + e_2)\varepsilon$, where $(e_1 + e_2)\varepsilon b\bar{\varepsilon}(e_1 + e_2)^{-1} \in \mathcal{O}$, because $e_1 + e_2$ is invariant. If a is a right divisor of c the assertion follows in an analogous way.

□

<u>Definition.</u> Given $a,b \in \mathcal{O} - \{0\}$ we say that a is a <u>total divisor</u> of b and write $a \| b$ if there exists $c \in I(\mathcal{O})$ such that $a |_l c |_l b$.

We observe that an element is not generally a total divisor of itself; in fact $a \| a$ if and only if a is invariant. In view of the preceding proposition one can equivalently demand $a |_r c |_r b$ in the definition.

By means of Lemma 1.11 we are also able to decompose an integral quaternion into irreducible factors.

Proposition 1.13. Suppose that $a \in \mathcal{O}(\mathbb{H})$ and $N(a) = \prod_{j=1}^{n} p_j$, where each p_j is a prime. Then there exist $a_j \in \mathcal{O}(\mathbb{H})$ such that $N(a_j) = p_j$, $1 \leq j \leq n$, and $a = a_1 \cdot \ldots \cdot a_n$.

Proof. Let $N(a) = qp$, where p is a prime. Suppose that there does not exist any $b \in \mathcal{O}(\mathbb{H})$ satisfying $N(b) = p$ and $b|_r a$. Since each prime is a sum of four squares, there always exists an integral quaternion b satisfying $N(b) = p$. Hence we can conclude that none of the elements $p\varepsilon$, $\varepsilon \in E$, is a right divisor of a. Thus p and a turn out to be relatively right-prime. Lemma 1.11 yields the existence of $\alpha, \beta \in \mathcal{O}(\mathbb{H})$ such that

$$\alpha a + \beta p = 1 .$$

Hence

$$N(\alpha a + \beta p) = N(\alpha)N(a) + p^2 N(\beta) + p\, 2\,\mathrm{Re}(\alpha a \bar{\beta}) .$$

Now p is a divisor of $N(\alpha a + \beta p) = 1$ and we have a contradiction. Therefore we can find a decomposition $a = cb$ satisfying $N(c) = q$ and $N(b) = p$. The assertion follows by induction on n. □

Let V be an $\mathbb{R}$-vector space of finite dimension together with a symmetric non-degenerate bilinear form μ. Given a lattice L in V then

$$L^{\mu} := \{v \in V \,;\, \mu(v,x) \in \mathbb{Z} \quad \text{for} \quad x \in L\}$$

turns out to be a lattice, too. One has $(L^{\mu})^{\mu} = L$ and calls L^{μ} the dual lattice of L with respect to μ.

According to Proposition 1.1 the map $\mathbb{F} \times \mathbb{F} \longrightarrow \mathbb{R}$, $(a,b) \longmapsto \mathrm{Re}(\bar{a}b)$, is a positive definite bilinear form on $\mathbb{F}$. We denote by $\mathcal{O}^{\#}$ the dual lattice of $\mathcal{O}$ with respect to this bilinear form.

Proposition 1.14. One has the following dual lattices of the orderings in $\mathbb{F}$: $\quad \mathcal{O}^{\#}(\mathbb{R}) = \mathcal{O}(\mathbb{R})$, $\mathcal{O}^{\#}(\mathbb{C}) = \mathcal{O}(\mathbb{C})$ and

$$\mathcal{O}^{\#}(\mathbb{H}) = \mathbb{Z}2e_1 + \mathbb{Z}(e_1+e_2) + \mathbb{Z}(e_1+e_3) + \mathbb{Z}(e_1+e_4) .$$

Proof. Proposition 1.1 implies $\mathrm{Re}(\bar{a}b) \in \mathbb{Z}$ for all $b \in \mathbb{Z}e_1 + \ldots + \mathbb{Z}e_r$ if and only if $a \in \mathbb{Z}e_1 + \ldots + \mathbb{Z}e_r$. Hence given $a = \sum_{j=1}^{4} a_j e_j \in \mathbb{H}$, where $a_j \in \mathbb{Z}$, $1 \leq j \leq 4$, one has $a \in \mathcal{O}^{\#}(\mathbb{H})$ if and only if $\frac{1}{2}\sum_{j=1}^{4} a_j = \mathrm{Re}(\bar{a}e_o) \in \mathbb{Z}$. Therefore

$$\mathcal{O}^{\#}(\mathbb{H}) = \{a = \sum_{j=1}^{4} a_j e_j \; ; \; a_j \in \mathbb{Z}, \; 1 \leq j \leq 4, \; \sum_{j=1}^{4} a_j \text{ even}\}$$

and this lattice can be represented in the form quoted above.

□

The elements of $\mathcal{O}^{\#}(\mathbb{H})$ are called <u>even quaternions</u>. We get another description by

Lemma 1.15. *The even quaternions form an ideal in* $\mathcal{O}(\mathbb{H})$ *and satisfy*

$$\mathcal{O}^{\#}(\mathbb{H}) = \{a \in \mathcal{O} \; ; \; N(a) \in 2\mathbb{Z}\}$$
$$= (e_1 + e_2)\mathcal{O} = \mathcal{O}(e_1 + e_2) \; .$$

Proof. Given $a = \sum_{j=1}^{4} a_j(e_1 + e_j)$, where $a_j \in \mathbb{Z}$, one easily computes $N(a) \in 2\mathbb{Z}$. Now consider $a \in \mathcal{O}$ with $N(a) \in 2\mathbb{Z}$. Then it follows that $a \in \mathbb{Z}e_1 + \ldots + \mathbb{Z}e_4$, because otherwise $a = \sum_{j=1}^{4}(a_j + \frac{1}{2})e_j$, where $a_j \in \mathbb{Z}$, would imply that $N(a) = 1 + \sum_{j=1}^{4}(a_j^2 + a_j) \notin 2\mathbb{Z}$. Hence $a = \sum_{j=1}^{4} a_j e_j$, where $a_j \in \mathbb{Z}$, and $N(a) = \sum_{j=1}^{4} a_j^2 \in 2\mathbb{Z}$ yields $\sum_{j=1}^{4} a_j \in 2\mathbb{Z}$ and $a \in \mathcal{O}^{\#}(\mathbb{H})$.

Since $e_1 + e_2$ is invariant, one has $(e_1+e_2)\mathcal{O} = \mathcal{O}(e_1+e_2) \subset \{a \in \mathcal{O} \; ; \; N(a) \in 2\mathbb{Z}\}$. One easily verifies $(e_1+e_j)(e_1+e_2)^{-1} \in \mathcal{O}$ for $1 \leq j \leq 4$ and thus completes the proof.

□

For further details, especially in view of factorization of quaternions of HURWITZ, the reader is referred to [25].

§2 Integral matrices

Depending on the special ordering the notions of integral and unimodular matrices are defined. The Euclidean algorithm enables us to derive a weak version of the Elementary divisor theorem, whereas questions of uniqueness are left out of account. Proofs of the case $\mathcal{O} = \mathbb{Z}$ are due to KOECHER [37], I,§1.

Given $m,n \in \mathbb{N}$ let $\mathrm{Mat}(m,n;\mathbb{F})$ denote the set of $m \times n$ matrices with coefficients in $\mathbb{F}$. We put $\mathrm{Mat}(n;\mathbb{F}) := \mathrm{Mat}(n,n;\mathbb{F})$ and denote the group of units in the ring $\mathrm{Mat}(n;\mathbb{F})$ by $GL(n;\mathbb{F})$. Using the notation $I = I^{(n)}$ for the $n \times n$ identity matrix then $GL(n;\mathbb{F})$ consists of all matrices $X \in \mathrm{Mat}(n;\mathbb{F})$ for which there exists $Y \in \mathrm{Mat}(n;\mathbb{F})$ such that $XY = YX = I$. The notion of the rank of a matrix can be transferred from fields to skew fields; for a more general situation the reader is referred to [14],5.5. Thus $\mathrm{Mat}(n;\mathbb{F})$ consists of all matrices $X \in \mathrm{Mat}(n;\mathbb{F})$ having rank n.

It is important to know certain matrix representations for the elements of $\mathbb{C}$ and $\mathbb{H}$. According to [56] or [27],2.4, we define the map

$$\hat{}: \mathbb{F} \longrightarrow \mathrm{Mat}(r;\mathbb{R}), \quad a \longmapsto \hat{a},$$

where $\hat{a}$ denotes the matrix of the endomorphism $\mathbb{F} \longrightarrow \mathbb{F}$, $x \longmapsto x\bar{a}$, with respect to the canonical basis $e_1,\dots,e_r$. More precisely, given a in the form $a = \sum_{j=1}^{r} a_j e_j$ this means

$$\hat{a} = (a) \qquad \text{for } \mathbb{F} = \mathbb{R},$$

$$\hat{a} = \begin{pmatrix} a_1 & a_2 \\ -a_2 & a_1 \end{pmatrix} \qquad \text{for } \mathbb{F} = \mathbb{C} \text{ and}$$

$$\hat{a} = \begin{pmatrix} a_1 & a_2 & a_3 & a_4 \\ -a_2 & a_1 & -a_4 & a_3 \\ -a_3 & a_4 & a_1 & -a_2 \\ -a_4 & -a_3 & a_2 & a_1 \end{pmatrix} \qquad \text{for } \mathbb{F} = \mathbb{H}.$$

In addition we define the map

$$^{\vee}: \mathbb{H} \longrightarrow \mathrm{Mat}(2;\mathbb{C}), \quad a \longmapsto \check{a} := \begin{pmatrix} a_1e_1+a_2e_2 & a_3e_1+a_4e_2 \\ -a_3e_1+a_4e_2 & a_1e_1-a_2e_2 \end{pmatrix}.$$

Given $a \in \mathbb{C}$ one has $\det \hat{a} = N(a)$ and $\hat{\bar{a}} = (\hat{a})'$. For $a \in \mathbb{H}$ one computes $\det \check{a} = N(a)$, $\overline{\check{a}} = \overline{(\check{a})'}$, $\det \hat{a} = N(a)^2$, $\hat{\bar{a}} = (\hat{a})'$.

Lemma 2.1. a) *The map*

$$\mathrm{Mat}(n;\mathbb{F}) \longrightarrow \mathrm{Mat}(rn;\mathbb{R}), \quad X = (x_{jk}) \longmapsto \hat{X} := (\hat{x}_{jk}),$$

where $\hat{x}_{jk}$ *denotes the* $r \times r$ *matrix related to* x_{jk}, *is an injective homomorphism of the* $\mathbb{R}$-*algebras satisfying* $\widehat{I^{(n)}} = I^{(rn)}$. *An* $n \times n$ *matrix* X *belongs to* $GL(n;\mathbb{F})$ *if and only if* $\hat{X}$ *lies in* $GL(4n;\mathbb{R})$. *One has* $\det \hat{X} = N(\det X)$ *for* $X \in \mathrm{Mat}(n;\mathbb{C})$.

b) *The similarly defined map*

$$\mathrm{Mat}(n;\mathbb{H}) \longrightarrow \mathrm{Mat}(2n;\mathbb{C}), \quad X = (x_{jk}) \longmapsto \check{X} := (\check{x}_{jk}),$$

turns out to be an injective homomorphism of the $\mathbb{R}$-*algebras satisfying* $\widehat{(\check{X})} = \hat{X}$ *and* $\widecheck{I^{(n)}} = I^{(2n)}$. *An* $n \times n$ *matrix* X *lies in* $GL(n;\mathbb{H})$ *if and only if* $\check{X}$ *belongs to* $GL(2n;\mathbb{C})$. *One has* $\det \hat{X} = (\det \check{X})^2$ *for* $X \in \mathrm{Mat}(n;\mathbb{H})$.

Proof. a) Obviously the map $X \longmapsto \hat{X}$ is $\mathbb{R}$-linear and injective. Because of $\widehat{ab} = \hat{a}\hat{b}$ for $a,b \in \mathbb{F}$ on has $\widehat{XY} = \hat{X}\hat{Y}$ for $X,Y \in \mathrm{Mat}(n;\mathbb{F})$. From $\widehat{I^{(n)}} = I^{(rn)}$ we get $\widehat{(X^{-1})} = (\hat{X})^{-1}$. If $X \in \mathrm{Mat}(n;\mathbb{F})$ is not invertible there exists $x \in \mathbb{F}^n - \{0\}$ such that $Xx = 0$. Hence $\hat{X}\hat{x} = 0$, where $0 \neq \hat{x} \in \mathrm{Mat}(rn,r;\mathbb{R})$. Thus $\hat{X}$ is not invertible. Now it suffices to prove the identity $\det \hat{X} = |\det X|^2$ for $X \in GL(n;\mathbb{C})$. This identity can easily be checked for triangular and permutation matrices. Since these matrices generate $GL(n;\mathbb{C})$, the assertion is true for all $X \in \mathrm{Mat}(n;\mathbb{C})$.

b) We use similar arguments. □

By means of the maps $X \longmapsto \check{X}$ resp. $X \longmapsto \hat{X}$ we get a substitute for the determinant of matrices in $\mathrm{Mat}(n;\mathbb{H})$.

Let $\mathcal{O} = \mathcal{O}(\mathbb{F})$ denote the ordering defined in §1. The elements of $\mathrm{Mat}(m,n;\mathcal{O})$, i.e. the $m \times n$ matrices with coefficients in $\mathcal{O}$, are called *integral matrices*. Together with $\mathcal{O}$ also $\mathrm{Mat}(n;\mathcal{O})$ turns out to be a unital ring. The set of *unimodular matrices*, i.e. the set of units in the ring $\mathrm{Mat}(n;\mathcal{O})$, is denoted by $GL(n;\mathcal{O})$. The *unimodular group* $GL(n;\mathcal{O})$ is a subgroup of $GL(n;\mathbb{F})$. Obviously $GL(1;\mathcal{O}) = \mathcal{E}(\mathbb{F})$.

We define I_{jk} to be the matrix in $\mathrm{Mat}(n;\mathbb{F})$ having a lone 1 as its (j,k)-entry and all other entries 0. The j-th unit vector in $\mathbb{F}^n$ is denoted by $\mathfrak{e}_j$. Now we state the main results of this paragraph:

<u>Theorem 2.2.</u> <u>Given</u> $n > 1$ <u>the unimodular group</u> $GL(n;\mathcal{O})$ <u>is generated by the matrices</u> $I + I_{1n}$, $[\varepsilon,1,\ldots,1]$, <u>where</u> $\varepsilon \in E$, <u>and the permutation matrices</u> $P_\pi := (\mathfrak{e}_{\pi(1)},\ldots,\mathfrak{e}_{\pi(n)})$, <u>where</u> $\pi \in \gamma_n$.

<u>Theorem 2.3.</u> <u>Elementary divisor theorem</u>
<u>Suppose that</u> $A \in \mathrm{Mat}(m,n;\mathcal{O})$, $q = \mathrm{rank}\, A > 0$. <u>There exist</u> $U \in GL(m;\mathcal{O})$, $V \in GL(n;\mathcal{O})$ <u>and a diagonal matrix</u> $D = [d_1,\ldots,d_q] \in \mathrm{Mat}(q;\mathcal{O})$ <u>such that</u> $d_1 \| d_2 \| \ldots \| d_q$ <u>and</u>

$$UAV = \begin{pmatrix} D & 0 \\ 0 & 0 \end{pmatrix} .$$

<u>Proof</u> of both theorems. Put $\Delta_1 := E(\mathbb{F})$. If $n > 1$, let Δ_n denote the subgroup of $GL(n;\mathcal{O})$ generated by the matrices $I + I_{1n}$, $[\varepsilon,1,\ldots,1]$, $\varepsilon \in E$, and P_π , $\pi \in \gamma_n$. Since all permutation matrices belong to Δ_n , we conclude that

(1) $\quad I + aI_{jk} \in \Delta_n$ for $j \neq k$ and $a \in \mathcal{O}$.

By induction on $n+m$ we prove that for any $A \in \mathrm{Mat}(m,n;\mathcal{O})$ satisfying $q = \mathrm{rank}\, A > 0$ there exist $U \in \Delta_m$, $V \in \Delta_n$ and $D = [d_1,\ldots,d_q] \in \mathrm{Mat}(q;\mathcal{O})$ such that $d_1 \| d_2 \| \ldots \| d_q$ and

$$UAV = \begin{pmatrix} D & 0 \\ 0 & 0 \end{pmatrix} .$$

The case $m+n = 2$ is trivial. Suppose that $m+n > 2$ and define

$$M = \{UAV ;\ U \in \Delta_m ,\ V \in \Delta_n\} ,$$
$$N = \{N(b);\ \text{there is } B = (b_{jk}) \in M ,\ \text{where } b = b_{jk} \neq 0\} .$$

We conclude $N \neq \emptyset$ from $A \neq 0$ and get the existence of $B \in M$ such that $N(b_{jk})$ is minimal in N . Since all permutation matrices belong to Δ_n , we can choose B such that $N(b_{11})$ is minimal. In the case $m > 1$ we multiply by matrices $I + aI_{j1}$, $j > 1$, $a \in \mathcal{O}$, from the left and by matrices $I + aI_{1j}$, $j > 1$, $a \in \mathcal{O}$, from the right. By means of the Euclidean algorithm (Corollary 1.8) and the minimality of $N(b_{11})$ we see that M contains a matrix of the form

$$B = [b_{11}, B_1] ,$$

where $B_1 \in \mathrm{Mat}(m-1,n-1;\mathcal{O})$, and the assertion is proved if $m = 1$ or

$n = 1$. If $B_1 = 0$ we are ready. Otherwise, by induction hypothesis there exist $U_1 \in \Delta_{m-1}$, $V_1 \in \Delta_{n-1}$ and $D_1 = [d_2,\dots,d_q] \in \mathrm{Mat}(q-1;\mathcal{O})$ such that $d_2 \| \dots \| d_q$ and

$$U_1 B_1 V_1 = \begin{pmatrix} D_1 & 0 \\ 0 & 0 \end{pmatrix} .$$

One easily verifies $U = [1,U_1] \in \Delta_m$ and $V = [1,V_1] \in \Delta_n$. Defining $D = [d_1,\dots,d_q]$, $d_1 = b_{11}$, we have

$$UBV = \begin{pmatrix} D & 0 \\ 0 & 0 \end{pmatrix} .$$

Now consider

$$\begin{pmatrix} 1 & x \\ 0 & 1 \end{pmatrix} \begin{pmatrix} d_1 & 0 \\ 0 & d_2 \end{pmatrix} = \begin{pmatrix} d_1 & xd_2 \\ 0 & d_2 \end{pmatrix} ,$$

where $x \in \mathcal{O}$. Hence by construction, d_1 is a left divisor of xd_2 and also of xd_2y , where $x,y \in \mathcal{O}$. In view of Proposition 1.9 the two-sided ideal generated by d_2 can be generated by an invariant element c . Therefore d_1 is a left divisor of c and thus a total divisor of d_2 .

In order to prove Theorem 2.2 consider $A \in GL(n;\mathcal{O})$. Choose $U,V \in \Delta_n$ and a diagonal matrix $D = [d_1,\dots,d_n]$ satisfying $UAV = D$. Since D is unimodular, $d_1,\dots,d_n$ belong to E . Now $D \in \Delta_n$ yields $A \in \Delta_n$. □

In the cases $\mathbb{F} = \mathbb{R},\mathbb{C}$ the diagonal elements are uniquely determined up to unit factors. If we consider the ring of the quaternions of HURWITZ the question of uniqueness turns out to be more complicated. Given $a,b \in \mathcal{O}$ such that $N(a) = N(b) = p$, where p is a prime, one can find unimodular matrices $U,V \in GL(2;\mathcal{O})$ satisfying

$$U \begin{pmatrix} 1 & 0 \\ 0 & a \end{pmatrix} V = \begin{pmatrix} 1 & 0 \\ 0 & b \end{pmatrix} .$$

If p however is sufficiently large one can find a,b such that

$$b \notin EaE .$$

For more details in a general context the reader is referred to [14] , Ch.8 .

The special case $n = 1$ is formulated in

<u>Corollary 2.4.</u> <u>Given</u> $a \in \mathcal{O}^m$ <u>there exist</u> $U \in GL(m;\mathcal{O})$ <u>and</u> $d \in \mathcal{O}$ <u>such that</u> $Ua = (d,0,\dots,0)'$.

An induction argument leads to

Corollary 2.5. Given $A \in Mat(n;\mathcal{O})$ *there exists* $U \in GL(n;\mathcal{O})$ *such that* UA *is an upper (resp. lower) triangular matrix.*

As a consequence of Lemma 2.1 we state

$$\det \hat{A} = N(\det A) \in \mathbb{Z} \quad \text{for} \quad A \in Mat(n;\mathcal{O}(\mathbb{C})) .$$

Considering the set of generators of $GL(n;\mathcal{O})$ stated in Theorem 2.2 and diagonal matrices we get

Corollary 2.6. Given $A \in Mat(n;\mathcal{O}(\mathbb{H}))$ *one has* $\det \check{A} \in \mathbb{Z}$ *and* $\det \hat{A} = (\det \check{A})^2$.

A characterization of unimodular matrices is given in

Theorem 2.7. Given $U \in Mat(n;\mathcal{O})$ *the following statements are equivalent:*

(i) $U \in GL(n;\mathcal{O})$.

(ii) $\overline{U}' \in GL(n;\mathcal{O})$.

(iii) $\det U \in \mathcal{E}(\mathbb{F})$ *for* $\mathbb{F} = \mathbb{R},\mathbb{C}$ *resp.* $\det \check{U} = 1$ *for* $\mathbb{F} = \mathbb{H}$.

(iv) *The map* $\mathcal{O}^n \longrightarrow \mathcal{O}^n$, $x \longmapsto Ux$, *is bijective.*

(v) *The map* $\mathcal{O}^n \longrightarrow \mathcal{O}^n$, $x \longmapsto Ux$, *is surjective.*

(vi) *There is* $V \in Mat(n;\mathcal{O})$ *such that* $UV = I$.

(vii) *There is* $V \in Mat(n;\mathcal{O})$ *such that* $VU = I$.

Proof. If $UV = I$ holds, one has $\hat{U},\hat{V} \in GL(rn;\mathbb{Q})$ satisfying $\hat{U}\hat{V} = I$. Hence, $\hat{V}\hat{U} = I$ and $VU = I$. Therefore the statements (i), (ii), (vi) and (vii) are equivalent.

The chain of implications "(i) $\Rightarrow$ (iv) $\Rightarrow$ (v) $\Rightarrow$ (vi)" is obvious.

Corollary 2.6 yields "(i) $\Rightarrow$ (iii)" . On the other hand suppose that there is $U \in Mat(n;\mathcal{O})$ satisfying (iii). According to Theorem 2.3 we can choose $V,W \in GL(n;\mathcal{O})$ and $D = [d_1,\dots,d_n] \in Mat(n;\mathcal{O})$ such that $VUW = D$. Hence, D satisfies (iii). Thus $d_1,\dots,d_n$ belong to $\mathcal{E}$ and D as well as U lie in $GL(n;\mathcal{O})$. □

If U belongs to $GL(n;\mathcal{O}(\mathbb{C}))$, the matrix $\overline{U}$ is also unimodular. For the quaternions of HURWITZ and $n \geq 2$ the same fact is not true any longer: Consider

$$U = \begin{pmatrix} e_1 & e_4 \\ e_3 & e_1+e_2 \end{pmatrix} \in \mathrm{Mat}(2;\mathcal{O}) .$$

From
$$\begin{pmatrix} e_1 & e_4 \\ e_3 & e_1+e_2 \end{pmatrix}\begin{pmatrix} e_1 & -e_4 \\ 0 & e_1 \end{pmatrix} = \begin{pmatrix} e_1 & 0 \\ e_3 & e_1 \end{pmatrix} \quad \text{and}$$

$$\begin{pmatrix} e_1 & -e_4 \\ -e_3 & e_1-e_2 \end{pmatrix}\begin{pmatrix} e_1 & e_4 \\ 0 & e_1 \end{pmatrix} = \begin{pmatrix} e_1 & 0 \\ -e_3 & e_2-2e_2 \end{pmatrix}$$

in connection with the preceding theorem, we conclude that U is unimodular, but $\bar{U}$ fails to be.

<u>Lemma 2.8.</u> *Let* $\mathfrak{a} \in \mathcal{O}^n$, $\mathfrak{a} \neq 0$.

a) *If* $U \in GL(n;\mathcal{O})$ *then* gcrd $\mathfrak{a}$ *and* gcrd $U\mathfrak{a}$ *coincide*.

b) $\mathfrak{a}$ *is the column of a unimodular matrix if and only if the elements of* $\mathfrak{a}$ *are relatively right-prime*.

<u>Proof.</u> a) Put $\mathfrak{a} = (a_1,\ldots,a_n)'$, $U = (u_{jk})$, $\mathfrak{b} = U\mathfrak{a} = (b_1,\ldots,b_n)'$. Let a denote a gcrd $\mathfrak{a}$ and b a gcrd $\mathfrak{b}$. Because of

$b_j = \sum_{k=1}^{n} u_{jk}a_k$ one has $a|_r b_j$, $1 \le j \le n$, in view of Proposition 1.10,

hence $a|_r b$. The same conclusion for U^{-1} instead of U yields $b|_r a$. Therefore a and b are left associated.

b) According to part a) and Corollary 2.4 we may suppose that $\mathfrak{a} = (a,0,\ldots,0)'$. If $\mathfrak{a}$ is relatively right-prime a is a unit and $\mathfrak{a}$ turns out to be the first column of aI . If $\mathfrak{a}$ is the first column of a unimodular matrix it follows from Theorem 2.7 (iii) that a is a unit. Hence the elements of $\mathfrak{a}$ are relatively right-prime.

□

We observe that in the quaternions of HURWITZ for $n > 1$ the columns of unimodular matrices are not relatively left-prime in general. But considering $\bar{U}'$ we can conclude that the rows of unimodular matrices are relatively left-prime.

§3 Hermitian matrices

In this paragraph the analysis of Hermitian matrices is examined. Special interest is devoted to positive definite matrices. Proofs are adapted from [37], I, §2.

Given a matrix $X = (x_{jk}) \in \mathrm{Mat}(n;\mathbb{F})$ we denote the conjugate matrix by $\bar{X} = (\bar{x}_{jk})$. Since the map $\mathbb{F} \longrightarrow \mathbb{F}$, $x \longmapsto \bar{x}$, is an involution, the map

$$\mathrm{Mat}(n;\mathbb{F}) \longrightarrow \mathrm{Mat}(n;\mathbb{F}) , \quad X \longmapsto \bar{X}' ,$$

becomes an involution, too. Those matrices X which are kept fixed under this involution, i.e. $X = \bar{X}'$, are called <u>Hermitian</u>. Put

$$\mathrm{Sym}(n;\mathbb{F}) := \{X \in \mathrm{Mat}(n;\mathbb{F}) ; X = \bar{X}'\} .$$

$\mathrm{Sym}(n;\mathbb{F})$ turns out to be a formal real JORDAN-algebra with respect to the symmetric product $X * Y := \frac{1}{2}(XY + YX)$. For further details the reader is referred to [9], Kap.XI.

Given a quadratic matrix $X = X^{(n)} = (x_{jk})$ the <u>trace</u> of X is defined by

$$\mathrm{tr}(X) := \mathrm{trace}(X) := \sum_{k=1}^{n} x_{kk} .$$

For matrices $A,B \in \mathrm{Mat}(m,n;\mathbb{F})$ we put

$$\tau(A,B) = \tfrac{1}{2}\, \mathrm{tr}(A\bar{B}' + B\bar{A}') .$$

Proceeding from $A = (a_{kl})$, where $a_{kl} = \sum_{j=1}^{r} a_{kl}^{(j)} e_j$, and from a corresponding form of B, we calculate

$$\tau(A,B) = \sum_{j=1}^{r} \sum_{k=1}^{m} \sum_{l=1}^{n} a_{kl}^{(j)}\, b_{kl}^{(j)} .$$

Hence τ turns out to be the canonical scalar product on $\mathrm{Mat}(m,n;\mathbb{F})$. Given $X,Y \in \mathrm{Sym}(n;\mathbb{F})$ it holds that

$$\tau(X,Y) = \tfrac{1}{2}\, \mathrm{tr}(XY + YX) = \mathrm{tr}(X * Y).$$

Thus one easily checks the

<u>Proposition 3.1.</u> $\mathrm{Sym}(n;\mathbb{F})$ <u>is an</u> $\mathbb{R}$-<u>vector space of dimension</u> $h = h(n;\mathbb{F}) = n + \frac{1}{2} rn(n-1)$ <u>and</u> τ <u>becomes a positive definite bilinear form on</u> $\mathrm{Sym}(n;\mathbb{F})$.

Given $A = A^{(n)}$ and $B = B^{(n,m)}$ one defines

$$A[B] := \bar{B}'AB .$$

If A is Hermitian, then $A[B]$ is Hermitian, too. Especially one has $X[x] \in \mathbb{R}$ for $X \in \mathrm{Sym}(n;\mathbb{F})$ and $x \in \mathbb{F}^n$.

An easy computation yields

<u>Lemma 3.2.</u> <u>Completion of squares</u>
<u>Given</u> $X = \begin{pmatrix} A & B \\ \bar{B}' & C \end{pmatrix} \in \mathrm{Sym}(n;\mathbb{F})$, <u>where</u> $A = \bar{A}' \in GL(m;\mathbb{F})$, <u>it holds that</u>

$$X\begin{bmatrix} I & -A^{-1}B \\ 0 & I \end{bmatrix} = \begin{pmatrix} A & 0 \\ 0 & C-A^{-1}[B] \end{pmatrix} .$$

<u>Lemma 3.3.</u> <u>Given</u> $X \in \mathrm{Sym}(n;\mathbb{F})$ <u>there are</u> $V \in GL(n;\mathbb{F})$ <u>satisfying</u> $V\bar{V}' = I$ <u>and a uniquely determined real diagonal matrix</u> $D = [d_1,\dots,d_n]$ <u>such that</u> $d_1 \leq \dots \leq d_n$ <u>and</u> $X[V] = D$.

For a <u>proof</u> the reader is referred to [23], Kap.III. On the other hand the proof can be adapted from the real symmetric resp. complex Hermitian case described for example in [38], Satz 6.2.5.

The diagonal elements $d_1,\dots,d_n$ are called the <u>eigenvalues</u> of X and we define

$$\det X := d_1 \cdot \dots \cdot d_n .$$

For $\mathbb{F} = \mathbb{R},\mathbb{C}$ this definition coincides with the usual notion of the determinant of a matrix.

Given a Hermitian matrix $X = (x_{kl})$ we always use the abbreviation $x_k := x_{kk}$ for the diagonal elements. The determinant of a Hermitian matrix is described more precisely in the following

<u>Theorem 3.4.</u> a) <u>Given</u> $X \in \mathrm{Sym}(n;\mathbb{F})$ <u>and</u> $a \in \mathbb{F}^n$ <u>then</u>

$$p(\lambda) := \det(X + \lambda a\bar{a}') , \quad \lambda \in \mathbb{R} ,$$

<u>is a polynomial in</u> λ <u>of degree</u> ≤ 1 .

b) <u>If</u> $X \in \mathrm{Sym}(n;\mathbb{F})$ <u>then</u> $\det X$ <u>is a polynomial in</u> $x_1,\dots,x_n,x_{kl}^{(j)}$, $1 \leq k < l \leq n$, $1 \leq j \leq r$, <u>having coefficients in</u> $\mathbb{Z}$.

Proof. a) We use induction on n, where $n = 1$ is trivial. Suppose that $n > 1$ and $a \neq 0$. Thus one easily checks that $\mathrm{rank}(a\bar{a}') = 1$. From Lemma 3.3 we get a matrix $V \in GL(n;\mathbb{F})$ satisfying $V\bar{V}' = I$ and an $a \in \mathbb{R}$, $a > 0$, such that $a\bar{a}'[V] = [a,0,\dots,0]$. Replacing X by $X[\bar{V}']$ we may assume that $a\bar{a}' = [a,0,\dots,0]$. We choose a representation

(1) $$X = \begin{pmatrix} x & \bar{x}' \\ x & Y \end{pmatrix},$$

where $x = x_1 \in \mathbb{R}$, $x \in \mathbb{F}^{n-1}$ and $Y \in \mathrm{Sym}(n-1;\mathbb{F})$. One calculates

(2) $$\begin{pmatrix} x+\lambda a & \bar{x}' \\ x & Y \end{pmatrix}\begin{bmatrix} 1 & -(x+\lambda a)^{-1}\,\bar{x}' \\ 0 & I \end{bmatrix} = \begin{pmatrix} x+\lambda a & 0 \\ 0 & Y-(x+\lambda a)^{-1}\,x\bar{x}' \end{pmatrix}.$$

for all $\lambda \in \mathbb{R}$ satisfying $x+\lambda a \neq 0$ according to Lemma 3.2. By induction hypothesis $\det(Y-(x+\lambda a)^{-1}x\bar{x}')$ is a polynomial in $(x+\lambda a)^{-1}$ of degree ≤ 1. Now $p(\lambda) = (x+\lambda a)\det(Y-(x+\lambda a)^{-1}x\bar{x}')$ holds for all $\lambda \in \mathbb{R}$ satisfying $x+\lambda a \neq 0$. Hence $p(\lambda)$ is a polynomial in λ of degree ≤ 1.

b) Since the case $n = 1$ is evident, we may assume that $n > 1$. We choose X in the form (1) and may suppose that $x \neq 0$ in view of an argument of continuity. Following (2) one has $\det X = x\det(Y-x^{-1}x\bar{x}')$. Hence part a) and the induction hypothesis yield the assertion.

□

Given a "small" value of n this polynomial can easily be calculated. If $X \in \mathrm{Sym}(2;\mathbb{F})$ one has

$$\det X = x_1x_2 - N(x_{12}).$$

In general $\det X$, $X \in \mathrm{Sym}(n;\mathbb{F})$, turns out to be the reduced norm of X in the sense of [9], Kap.XI.

Corollary 3.5. a) *Given* $X \in \mathrm{Sym}(n;\mathbb{F})$ *one has* $\det \hat{X} = (\det X)^r$. *In addition* $\det \check{X} = (\det X)^2$ *holds for* $X \in \mathrm{Sym}(n;\mathbb{H})$.

b) *If* $X \in \mathrm{Sym}(n;\mathbb{F})$ *is integral it follows that*

$$\det X \in \mathbb{Z}.$$

Proof. a) Use Lemma 3.3 and consider $\hat{D}$ resp. $\check{D}$, where D is a real diagonal matrix.

b) The assertion is obvious for $\mathbb{F} = \mathbb{R},\mathbb{C}$. Considering $\mathbb{F} = \mathbb{H}$ we have $\det X \in \mathbb{Q}$ in view of Theorem 3.4. But Corollary 2.6 yields $(\det X)^2 = \det \check{X} \in \mathbb{Z}$, hence $\det X \in \mathbb{Z}$. □

A special class of Hermitian matrices is to be examined more precisely.

Definition. A Hermitian matrix $X \in Sym(n;\mathbb{F})$ is called positive definite resp. positive semi-definite if every $x \in \mathbb{F}^n - \{0\}$ fulfills $X[x] > 0$ resp. $X[x] \geq 0$. We use the notations $X > 0$ resp. $X \geq 0$ and

$$Pos(n;\mathbb{F}) := \{X \in Sym(n;\mathbb{F})\ ;\ X > 0\} .$$

Given $X = (x_{kl}) \in Sym(n;\mathbb{F})$ and $y = (y_1,\ldots,y_n)' \in \mathbb{F}^n$ one calculates

$$\text{(3)} \qquad X[y] = \sum_{k=1}^{n} x_k\, N(y_k) + \sum_{1 \leq k < l \leq n} 2\, Re(\bar{y}_k x_{kl} y_l) .$$

Together with the obvious relation

$$Sym(n;\mathbb{R}) \subset Sym(n;\mathbb{C}) \subset Sym(n;\mathbb{H})$$

(3) leads to

$$Pos(n;\mathbb{R}) \subset Pos(n;\mathbb{C}) \subset Pos(n;\mathbb{H}) .$$

Given $A \in Mat(n;\mathbb{F})$ one has $\widehat{(\bar{A}')} = (\hat{A})'$ and for $\mathbb{F} = \mathbb{H}$ additionally $(\check{\bar{A}}') = \overline{(\check{A})}'$. Thus

$$\{\hat{X}\ ;\ X \in Sym(n;\mathbb{F})\} \subset Sym(rn;\mathbb{R})$$

and

$$\{\check{X}\ ;\ X \in Sym(n;\mathbb{H})\} \subset Sym(2n;\mathbb{C}) .$$

More precisely, we observe that on the one hand $Sym(n;\mathbb{R})$ is a real submanifold of $Sym(n;\mathbb{H})$ and that on the other hand $Sym(n;\mathbb{H})$ can be regarded as a submanifold of $Sym(4n;\mathbb{R})$ via $X \longmapsto \hat{X}$.

We summarize some equivalent assertions concerning Hermitian matrices.

Theorem 3.6. Given $X \in Sym(n;\mathbb{F})$ the following assertions are equivalent:

(i) X is positive definite.

(ii) $\hat{X} \in Pos(rn;\mathbb{R})$ or $\check{X} \in Pos(2n;\mathbb{C})$ for $\mathbb{F} = \mathbb{H}$.

(iii) There is $W \in GL(n;\mathbb{F})$ such that $X = \bar{W}'W$.

(iv) X is positive semi-definite and invertible.

(v) The inverse X^{-1} exists and is positive definite.

(vi) Given $A \in GL(n;\mathbb{F})$ one has $X[A] > 0$.

(vii) *There exist* $V \in GL(n;\mathbb{F})$ *satisfying* $V\bar{V}' = I$ *and a diagonal matrix* D *consisting of real positive elements such that* $X[V] = D$.

Proof. The crucial equivalence "(i) ⇔ (vii)" follows from Lemma 3.3. The other statements arise by simple conclusions, where we have to bear in mind that $X[x]I^{(r)} = \widehat{X[x]} = \hat{X}[\hat{x}]$ holds for $x \in \mathbb{F}^n$.

□

Arguments of continuity lead to

Corollary 3.7. *Given* $X \in Sym(n;\mathbb{F})$ *the following assertions are equivalent*:

(i) X *is positive semi-definite.*

(ii) $\hat{X} \geq 0$ *or* $\check{X} \geq 0$ *for* $\mathbb{F} = \mathbb{H}$.

(iii) *There is* $W \in Mat(n;\mathbb{F})$ *such that* $X = \bar{W}'W$.

(iv) $X[A] \geq 0$ *for all* $A \in Mat(n,m;\mathbb{F})$.

(v) *There exist* $V \in GL(n;\mathbb{F})$ *satisfying* $V\bar{V}' = I$ *and a diagonal matrix* D *consisting of real non-negative elements such that* $X[V] = D$.

One easily adapts the proofs for real symmetric matrices in order to demonstrate

Corollary 3.8. *Suppose that* $S,X,Y \in Sym(n;\mathbb{F})$, *where* S *is arbitrary*, $0 \neq X$ *is positive semi-definite and* Y *is positive definite*.

a) *The diagonal elements of* X *are non-negative and* $tr\ X > 0$.

b) *The diagonal elements of* Y , $\det Y$ *and* $\tau(X,Y) = \frac{1}{2}\, tr(XY+YX)$ *are positive*.

c) *There exist* $W \in GL(n;\mathbb{F})$ *and a real diagonal matrix* D *such that* $Y[W] = I$ *and* $S[W] = D$.

The following statement is known as HADAMARD'*s inequality*.

Corollary 3.9. *Given* $X \in Pos(n;\mathbb{F})$ *it holds that*

$$\det X \leq x_1 \cdot \ldots \cdot x_n .$$

Proof. We choose a decomposition $X = \begin{pmatrix} Y & x \\ \bar{x}' & x \end{pmatrix}$, where $Y \in Pos(n-1;\mathbb{F})$ $x \in \mathbb{F}^{n-1}$ and $x = x_n$. Lemma 3.2 yields

$$\det X = (x - Y^{-1}[x]) \det Y .$$

From $X > 0$ we get $0 < x - Y^{-1}[x] \leq x$, hence $\det X \leq x \det Y$. An induction on n completes the proof. □

Corollary 3.10. JACOBIan representation

Every $X \in \mathrm{Pos}(n;\mathbb{F})$ possesses a unique representation $X = D[B]$, where $D > 0$ is a diagonal matrix and $B \in \mathrm{Mat}(n;\mathbb{F})$ is an upper triangular matrix having only 1's as diagonal entries.

Proof. The existence follows from Lemma 3.2 by induction on n. Let $X = D[B] = F[G]$ denote two representations of this kind. Since BG^{-1} is also an upper triangular matrix possessing only 1's as diagonal entries, $D[BG^{-1}] = F$ implies $BG^{-1} = I$, hence $B = G$ and $D = F$. □

Given $X \in \mathrm{Sym}(n;\mathbb{F})$ and $1 \leq k \leq n$ we put

$$X_k := X\begin{bmatrix} I \\ 0 \end{bmatrix} , \text{ where } I = I^{(k)} .$$

One has $X_k \in \mathrm{Sym}(k;\mathbb{F})$ and calls $d_k(X) := \det X_k$ the k-th principal minor of X. Especially, $d_1(X) = x_1$ and $d_n(X) = \det X$.

Theorem 3.11. A matrix $X \in \mathrm{Sym}(n;\mathbb{F})$ is positive definite if and only if all principal minors $d_1(X),\dots,d_n(X)$ are positive.

Proof. From $X > 0$ we conclude $X_k > 0$. Corollary 3.8 yields $\det X_k > 0$. The other implication is proved by induction on n, where $n = 1$ is trivial. Given $n > 1$ we choose a decomposition

$$X = \begin{pmatrix} X_{n-1} & x \\ \bar{x}' & x_n \end{pmatrix} , \quad x \in \mathbb{F}^{n-1} .$$

The induction hypothesis yields $X_{n-1} > 0$, hence $X_{n-1} \in GL(n-1;\mathbb{F})$. The application of Lemma 3.2 leads to

$$Y = X\begin{bmatrix} I & -X_{n-1}^{-1}x \\ 0 & 1 \end{bmatrix} = \begin{pmatrix} X_{n-1} & 0 \\ 0 & z \end{pmatrix} , \quad z = x_n - X_{n-1}^{-1}[x] ,$$

hence $\det X = (x_n - X_{n-1}^{-1}[x]) \det X_{n-1}$. Thus $x_n - X_{n-1}^{-1}[x] > 0$, and $X_{n-1} > 0$ implies $Y > 0$ as well as $X > 0$. □

The proof of the following assertion is obvious.

Proposition 3.12. $Pos(n;\mathbb{F})$ is an open subset of $Sym(n;\mathbb{F})$ possessing $\overline{Pos(n;\mathbb{F})} = \{X \in Sym(n;\mathbb{F})\ ;\ X \geq 0\}$ as its topological closure. The boundary consists of the non-invertible positive semi-definite matrices. $Pos(n;\mathbb{F})$ and $\overline{Pos(n;\mathbb{F})}$ are convex cones.

Given $S,T \in Sym(n;\mathbb{F})$ we define as usual

$$S > T \iff S - T > 0\ ,$$
$$S \geq T \iff S - T \geq 0\ .$$

The relations ">" and "≥" define partial orderings on $Sym(n;\mathbb{F})$. We can derive statements on these partial orderings from properties of $Pos(n;\mathbb{F})$.

Lemma 3.13. Let C be a compact subset of $Pos(n;\mathbb{F})$. Then one can find $\alpha,\beta \in \mathbb{R}$, $\alpha \geq \beta > 0$, such that

$$\alpha I \geq T \geq \beta I \quad \text{for every} \quad T \in C\ .$$

Proof. $P := C \times \{x \in \mathbb{F}^n\ ;\ \bar{x}'x = 1\}$ is a compact subset of $Pos(n;\mathbb{F}) \times \mathbb{F}^n$. The map $P \longrightarrow \mathbb{R}$, $(T,x) \longmapsto T[x]$, is continuous; let α denote its maximum and β its minimum. □

Lemma 3.14. Given $S \in Pos(n;\mathbb{F})$ and $T \in Sym(m;\mathbb{F})$ the set $\{A \in Mat(n,m;\mathbb{F})\ ;\ T \geq S[A]\}$ is bounded.

Proof. According to the preceding lemma there exists $\alpha > 0$ such that $S \geq \alpha I$. Given A in the form (a_{kl}) then $S[A] \geq \alpha\bar{A}'A$ and Corollary 3.8 yields

$$\mathrm{tr}\ T \geq \mathrm{tr}\ S[A] \geq \alpha\ \mathrm{tr}\ \bar{A}'A = \alpha \sum_{k=1}^{n} \sum_{l=1}^{m} N(a_{kl})\ .$$ □

Suppose that $S = (s_{jk}) \in Sym(n;\mathbb{F})$. We define the diagonal of S by

$$\mathrm{diag}\ S := [s_1,\ldots,s_n]\ .$$

Lemma 3.15. Suppose that $n > 1$ and $S \in Sym(n;\mathbb{F})$.

a) If S is positive definite then $n\,\mathrm{diag}\,S > S$ and $(\mathrm{tr}\ S)I > S$.

b) If S is positive semi-definite then $n\,\mathrm{diag}\,S \geq S$ and $(\mathrm{tr}\ S)I \geq S$.

<u>Proof.</u> a) According to Theorem 3.6 there exist $V \in GL(n;\mathbb{F})$ and a diagonal matrix D such that $V\bar{V} = I$ and $S[V] = D$. Because of $\mathrm{tr}\, S = \mathrm{tr}\, D$ one has $(\mathrm{tr}\, S)I > S$ if and only if $(\mathrm{tr}\, D)I > D$. The last assertion is obvious. As $\mathrm{diag}\, S$ is positive definite, there exists a diagonal matrix F such that $F^2 = \mathrm{diag}\, S$. Consider $T := S[F^{-1}]$. The observation $\mathrm{diag}\, T = I$ leads to $nI > T$, hence to

$$n\, \mathrm{diag}\, S = n\, F^2 > T[F] = S\ .$$

b) Use an argument of continuity. □

Since the case $n = 1$ is trivial, we apply the preceding assertions to the case $n = 2$. Given $S = \begin{pmatrix} s_1 & s \\ s & s_2 \end{pmatrix} \in \mathrm{Sym}(2;\mathbb{F})$ Theorem 3.11 leads to

$$S > 0 \iff s_1 > 0 \text{ and } \det S = s_1 s_2 - N(s) > 0\ ,$$

$$S \geq 0 \iff s_1 \geq 0\ ,\ s_2 \geq 0 \text{ and } \det S = s_1 s_2 - N(s) \geq 0\ .$$

If S is positive definite the JACOBIan representation of S is obviously given by $S = D[B]$, where

$$D = \begin{pmatrix} s_1 & 0 \\ 0 & s_2 - s_1^{-1} N(s) \end{pmatrix}\ ,\quad B = \begin{pmatrix} 1 & s_1^{-1} s \\ 0 & 1 \end{pmatrix}\ .$$

The eigenvalues of $S \in \mathrm{Sym}(2;\mathbb{F})$ turn out to be the zeros of the real "characteristic" polynomial

$$X^2 - \mathrm{tr}(S)X + \det S \in \mathbb{R}[X]\ .$$

§4 MINKOWSKI's reduction theory

In this paragraph the Main theorem of MINKOWSKI's reduction theory is derived. WEYL [61] applied MINKOWSKI's methods to Hermitian forms over the quaternions of HURWITZ. By other means KOECHER [36] gave a version of the main theorem in the more general context of domains of positivity. Statements and proofs of the real symmetric case are due to CHRISTIAN [13], IV, §1, FREITAG [21], I, §2, II, §1, KOECHER [37], I, §3, §5, and MAASS [45], §9.

Two matrices $S, T \in Sym(n;\mathbb{F})$ are called <u>integrally equivalent</u> if there is $U \in GL(n;\mathcal{O})$ satisfying $S[U] = T$. Thus an equivalence relation is defined on $Sym(n;\mathbb{F})$. Given $S \in Sym(n;\mathbb{F})$ then $\det S$ depends only on the equivalence class of S. By $Y \longmapsto Y[W]$, $W \in GL(n;\mathcal{O})$, the cone $Pos(n;\mathbb{F})$ is mapped bijectively onto itself. Hence the whole equivalence class of any $S \in Pos(n;\mathbb{F})$ lies in $Pos(n;\mathbb{F})$.

Given $S \in Pos(n;\mathbb{F})$ we define

$$\mu(S) := \mu(S;\mathbb{F}) := \inf \{S[\mathfrak{g}] \; ; \; \mathfrak{g} \in \mathcal{O}^n - \{0\}\}.$$

$\mu(S)$ is called the <u>minimum</u> of S.

<u>Proposition 4.1.</u> <u>Given</u> $S \in Pos(n;\mathbb{F})$ <u>there exists</u> $0 \neq \mathfrak{g} \in \mathcal{O}^n$ <u>such that</u> $\mu(S) = S[\mathfrak{g}] > 0$ <u>and</u> $\mu(S) = \mu(S[U])$ <u>holds for</u> $U \in GL(n;\mathcal{O})$. <u>From</u> $S > T > 0$ <u>the relation</u> $\mu(S) > \mu(T)$ <u>follows.</u>

<u>Proof.</u> The set $M := \{\mathfrak{g} \in \mathcal{O}^n \; ; \; \mathfrak{g} \neq 0 \; , \; S[\mathfrak{e}_1] \geq S[\mathfrak{g}]\}$ is bounded in view of Lemma 3.14. Since $\mathcal{O}$ is a discrete subset of $\mathbb{F}$, the set M is finite. Hence we can find $0 \neq \mathfrak{g} \in \mathcal{O}^n$ satisfying $\mu(S) = S[\mathfrak{g}] > 0$. From $S > T$ we conclude that $\mu(S) = S[\mathfrak{g}] > T[\mathfrak{g}] \geq \mu(T)$. The identity $\mu(S) = \mu(S[U])$, $U \in GL(n;\mathcal{O})$, follows from Theorem 2.7 (iv). □

<u>Corollary 4.2.</u> <u>Given</u> $S \in Pos(n;\mathbb{F})$ <u>there is</u> $U \in GL(n;\mathcal{O})$ <u>such that</u> $S[U] = T = (t_{kl})$ <u>and</u> $t_1 = \mu(S) = \mu(T)$.

<u>Proof.</u> Choose $0 \neq \mathfrak{g} \in \mathcal{O}^n$ such that $\mu(S) = S[\mathfrak{g}]$ according to the preceding proposition. Since $S[\mathfrak{g}]$ is minimal, $\mathfrak{g}$ is relatively right-prime. According to Lemma 2.8 $\mathfrak{g}$ is the first column of a matrix $U \in GL(n;\mathcal{O})$. □

A relation between two invariants of the equivalence relation is given by

Theorem 4.3. HERMITE's inequality
There is $d = d(\mathbb{F}) > 0$ such that

$$\mu(S;\mathbb{F})^n \leq d^{\,n(n-1)/2} \det S$$

holds for all $S \in \mathrm{Pos}(n;\mathbb{F})$. One can choose $d(\mathbb{R}) = \frac{4}{3}$ and $d(\mathbb{C}) = d(\mathbb{H}) = 2$.

Proof. The assertion is proved by induction on n , where $n = 1$ is obvious. Given $n > 1$ we may assume $s_1 = \mu(S)$ in view of the corollary above. According to Lemma 3.2 there exists a decomposition

$$S = \begin{pmatrix} s_1 & 0 \\ 0 & T \end{pmatrix}\begin{bmatrix} 1 & \bar{a}' \\ 0 & I \end{bmatrix} = \begin{pmatrix} s_1 & s_1\bar{a}' \\ s_1 a & T+s_1 a\bar{a}' \end{pmatrix},$$

where $T \in \mathrm{Pos}(n-1;\mathbb{F})$, $a \in \mathbb{F}^{n-1}$. Now choose $\mathfrak{g} \in \mathcal{O}^{n-1}$ satisfying $\mu(T) = T[\mathfrak{g}]$. Using Corollary 1.8 one can find $g \in \mathcal{O}$ such that $N(g+\bar{a}'\mathfrak{g}) \leq c = c(\mathbb{F})$. Hence

$$\mu(S) \leq S\begin{bmatrix} g \\ \mathfrak{g} \end{bmatrix} = s_1 N(g+\bar{a}'\mathfrak{g}) + T[\mathfrak{g}] \leq s_1 c + \mu(T) .$$

Defining $d := (1-c)^{-1}$ one has $\mu(S) \leq d\,\mu(T)$. The induction hypothesis applied to T yields

$$\mu(S)^n \leq \mu(S)d^{n-1}\mu(T)^{n-1} \leq \mu(S)d^{n-1}d^{(n-1)(n-2)/2}\det T = d^{n(n-1)/2}\det S . \qquad \square$$

Now we are going to define the domain of reduced matrices in $\mathrm{Pos}(n;\mathbb{F})$. Therefore we remember the definitions of $F(\mathbb{F})$ resp. $F^*(\mathbb{F})$ in §1.

By $R(n;\mathbb{F})$ we mean the set of matrices $S = (s_{kl}) \in \mathrm{Pos}(n;\mathbb{F})$ satisfying the following conditions :

(R.1) $s_{12} \in F^*(\mathbb{F})$ and $s_{1k} \in F(\mathbb{F})$ for $2 < k \leq n$,

(R.2) $S[\mathfrak{g}] \geq s_k$ for each vector $\mathfrak{g} = (g_1,\ldots,g_n)' \in \mathcal{O}^n$, where $g_k,\ldots,g_n$ are relatively right-prime, and for $1 \leq k \leq n$.

(R.1) and (R.2) are called MINKOWSKI's conditions of reduction. The elements of $R(n;\mathbb{F})$ are said to be reduced matrices. We use the abbreviation k-primitive for vectors $\mathfrak{g} = (g_1,\ldots,g_n)' \in \mathcal{O}^n$ such that $g_k,\ldots,g_n$ are relatively right-prime.

The "Main theorem of reduction theory" is deduced in several steps.

<u>Proposition 4.4.</u> <u>Given</u> $S \in \mathrm{Pos}(n;\mathbb{F})$ <u>there exists</u> $U \in GL(n;\mathcal{O})$ <u>such that</u>

$$S[U] \in \mathcal{R}(n;\mathbb{F}) .$$

<u>Proof.</u> Write U in the form $U = (\mathfrak{u}_1,\dots,\mathfrak{u}_n)$, where $\mathfrak{u}_j \in \mathcal{O}^n$. Choose $\mathfrak{u}_1 \in \mathcal{O}^n$ such that $\mu(S) = S[\mathfrak{u}_1]$. Thus $\mathfrak{u}_1$ is relatively right-prime. Now choose $\mathfrak{u}_2$ among all second columns of unimodular matrices having $\mathfrak{u}_1$ as its first column such that $S[\mathfrak{u}_2]$ becomes minimal. Replacing $\mathfrak{u}_1$ resp. $\mathfrak{u}_2$ by $\mathfrak{u}_1\varepsilon$ resp. $\mathfrak{u}_2\delta$, where $\varepsilon,\delta \in E(\mathbb{F})$, we may suppose according to Theorem 1.4 that $\bar{\mathfrak{u}}_1'S\mathfrak{u}_2 \in F^*(\mathbb{F})$ holds. Subsequently choose $\mathfrak{u}_3$ among all third columns of unimodular matrices $(\mathfrak{u}_1,\mathfrak{u}_2,*,\dots,*)$ such that $S[\mathfrak{u}_3]$ becomes minimal. We may replace $\mathfrak{u}_3$ by $\mathfrak{u}_3\rho$, $\rho \in E(\mathbb{F})$, in order to get $\bar{\mathfrak{u}}_1'S\mathfrak{u}_3 \in F(\mathbb{F})$. By continuation of this procedure we construct a unimodular matrix U and define $T := S[U]$.

The condition (R.1) holds because of the described construction. Given a k-primitive vector $\mathfrak{g} = (g_1,\dots,g_n)' \in \mathcal{O}^n$ there exists a unimodular matrix V having $(g_k,\dots,g_n)'$ as its first column according to Lemma 2.8. Put $G \in \mathrm{Mat}(k-1,n-k+1;\mathcal{O})$ possessing $(g_1,\dots,g_{k-1})'$ as its first column. Hence $W := U \begin{pmatrix} I & G \\ 0 & V \end{pmatrix} = (\mathfrak{u}_1,\dots,\mathfrak{u}_{k-1},\mathfrak{w},*,\dots,*) \in GL(n;\mathcal{O})$. By construction we have

$$T[\mathfrak{g}] = S[U\mathfrak{g}] = S[\mathfrak{w}] \geq S[\mathfrak{u}_k] = t_k .$$

Thus T is reduced. □

<u>Proposition 4.5.</u> $\mathcal{R}(n;\mathbb{F})$ <u>is closed in</u> $\mathrm{Pos}(n;\mathbb{F})$ <u>and contains interior points</u>.

<u>Proof.</u> Since $F(\mathbb{F})$ and $F^*(\mathbb{F})$ are closed in $\mathbb{F}$, it follows from the definition that $\mathcal{R}(n;\mathbb{F})$ is closed in $\mathrm{Pos}(n;\mathbb{F})$. According to the preceding proposition $\mathrm{Pos}(n;\mathbb{F})$ is a countable union of closed subsets:

$$\mathrm{Pos}(n;\mathbb{F}) = \bigcup_U \mathcal{R}(n;\mathbb{F})[U] , \text{ where } U \in GL(n;\mathcal{O}) .$$

Hence at least one set $\mathcal{R}(n;\mathbb{F})[U]$ and thus $\mathcal{R}(n;\mathbb{F})$ contain interior points. For a proof of the last fact the reader is referred to [22] , II.3.1 . □

Let $n > 1$. A transformation $Pos(n;\mathbb{F}) \longrightarrow Pos(n;\mathbb{F})$, $S \longmapsto S[U]$, where $U \in GL(n;\mathcal{O})$, is the identical map if and only if $U = \varepsilon I$, where $\varepsilon \in \mathcal{S}(\mathbb{F}) = \text{Cent}\,\mathbb{F} \cap \mathcal{E}(\mathbb{F})$.

<u>Proposition 4.6.</u> <u>Suppose that</u> $n > 1$, $S,T \in \mathcal{R}(n;\mathbb{F})$ <u>and</u> $U \in GL(n;\mathcal{O})$, $U \neq \varepsilon I$, $\varepsilon \in \mathcal{S}(\mathbb{F})$, <u>satisfying</u> $S[U] = T$. <u>Then</u> S <u>and</u> T <u>belong to the boundary of</u> $\mathcal{R}(n;\mathbb{F})$.

<u>Proof.</u> First assume that $U = [\varepsilon_1,\ldots,\varepsilon_n]$ is a diagonal matrix. Then $t_{1k} = \bar{\varepsilon}_1 s_{1k}\varepsilon_k$ holds for $1 < k \leq n$. Because of $s_{12},t_{12} \in \mathcal{F}^*(\mathbb{F})$ and $s_{1k},t_{1k} \in \mathcal{F}(\mathbb{F})$ for $2 < k \leq n$ at least one pair s_{1k},t_{1k} belongs to the boundary in view of Theorem 1.4.

If $U = (\mathfrak{u}_1,\ldots,\mathfrak{u}_n)$ is not a diagonal matrix choose k minimal such that $\mathfrak{u}_k \neq \varepsilon\mathfrak{e}_k$, $\varepsilon \in \mathcal{E}(\mathbb{F})$. Put $\mathfrak{u}_k = (u_1,\ldots,u_n)'$. Since U possesses the form $U = (\varepsilon_1\mathfrak{e}_1,\ldots,\varepsilon_{k-1}\mathfrak{e}_{k-1},\mathfrak{u}_k,*,\ldots,*)$, we observe that $(u_k,\ldots,u_n)'$ is the first column of a unimodular matrix. In view of Lemma 2.8 $\mathfrak{u}_k$ turns out to be k-primitive. On the other hand, U^{-1} has the form $U^{-1} = (\bar{\varepsilon}_1\mathfrak{e}_1,\ldots,\bar{\varepsilon}_{k-1}\mathfrak{e}_{k-1},\mathfrak{v},*,\ldots,*)$, where $\mathfrak{v}$ is k-primitve, too. Condition (R.2) applied to S and T yields $s_k = T[\mathfrak{v}] \geq t_k$ and $t_k = S[\mathfrak{u}_k] \geq s_k$, hence $s_k = t_k$. From $\mathfrak{u}_k,\mathfrak{v} \neq \varepsilon\mathfrak{e}_k$, $\varepsilon \in \mathcal{E}$, we conclude that S and T belong to the boundary of $\mathcal{R}(n;\mathbb{F})$.

□

In order to derive the conditions (F.4) and (F.5) we have to state some properties of reduced matrices.

<u>Proposition 4.7.</u> <u>Let</u> $S \in \mathcal{R}(n;\mathbb{F})$.

a) $s_1 = \mu(S;\mathbb{F})$.

b) $s_1 \leq \ldots \leq s_n$.

c) <u>For</u> $1 \leq k < l \leq n$ <u>and</u> $s_{kl} = \sum_{j=1}^{r} s_{kl}^{(j)} e_j$ <u>one has in the case</u>

$\mathbb{F} = \mathbb{R}$: $|s_{kl}| \leq \frac{1}{2}s_k$,

$\mathbb{F} = \mathbb{C}$: $|s_{kl}^{(j)}| \leq \frac{1}{2}s_k$, $N(s_{kl}) \leq \frac{1}{2}s_k^2$,

$\mathbb{F} = \mathbb{H}$: $|s_{kl}^{(j)}| \leq \frac{1}{2}s_k$, $\sum_{j=1}^{4} |s_{kl}^{(j)}| \leq s_k$, $N(s_{kl}) \leq \frac{1}{2}s_k^2$.

<u>Proof.</u> Given $\mathfrak{g} \in \mathcal{O}^n$ satisfying $S[\mathfrak{g}] = \mu(S)$ then $\mathfrak{g}$ is relatively right-prime. As S is reduced, $s_1 \leq S[\mathfrak{g}]$ holds and therefore

$s_1 = \mu(S)$. For $1 \leq k < n$ the unit vector $\mathfrak{e}_{k+1}$ becomes k-primitive, hence $s_k \leq S[\mathfrak{e}_{k+1}] = s_{k+1}$.

Given $1 \leq k < l \leq n$ and $\varepsilon \in E$ the vector $\mathfrak{g} = -\varepsilon\mathfrak{e}_k + \mathfrak{e}_l$ turns out to be l-primitive, hence

$$s_l \leq S[\mathfrak{g}] = s_k + s_l - 2\,\mathrm{Re}\,(\bar{\varepsilon}s_{kl}) \ , \ \mathrm{Re}(\bar{\varepsilon}s_{kl}) \leq \tfrac{1}{2}s_k \ .$$

Considering all units yields part c), where we have to bear in mind the proof of Corollary 1.7 in the case $\mathbb{F} = \mathbb{H}$. □

Now we are going to prove the crucial

Proposition 4.8. *There is a constant* $c_n = c(n;\mathbb{F}) > 0$ *such that*

$$s_1 \cdot \ldots \cdot s_n \leq c_n \det S$$

for every $S \in R(n;\mathbb{F})$.

Proof. We may choose $c_1 = 1$ and suppose that $c_1,\ldots,c_{n-1}$ have already been constructed. Conversely, we now assume that there exists a sequence of matrices S in $R(n;\mathbb{F})$ satisfying

(1) $$\frac{s_1 \cdot \ldots \cdot s_n}{\det S} \longrightarrow \infty \ .$$

If for every k , $1 \leq k < n$, the sequence of quotients $\frac{s_{k+1}}{s_k}$ were bounded, there would exist a constant c such that $s_1 \cdot \ldots \cdot s_n \leq c s_1^n$ for all S . Applying HERMITE's inequality (Theorem 4.3) this contradicts (1).

Hence we can choose k , $1 \leq k < n$, maximal such that the sequence $\frac{s_{k+1}}{s_k}$ is not bounded. Since the sequences $\frac{s_{k+2}}{s_{k+1}},\ldots,\frac{s_n}{s_{n-1}}$ are bounded, we can find $c > 0$ satisfying

(2) $$s_{k+1} \cdot \ldots \cdot s_n \leq c\, s_{k+1}^{n-k} \quad \text{for all } S \ .$$

As the sequence $\frac{s_{k+1}}{s_k}$ is not bounded, after passing to a subsequence we may assume that

(3) $$\frac{s_{k+1}}{s_k} \geq k^2 \quad \text{for all } S \ .$$

According to Lemma 3.2 we find a decomposition

$$S = \begin{pmatrix} S_1 & 0 \\ 0 & S_2 \end{pmatrix} \begin{bmatrix} I & B \\ 0 & I \end{bmatrix} , \text{ where } B = B^{(k,n-k)} \ .$$

Given $\mathfrak{g} = \begin{pmatrix}\mathfrak{g}_1\\ \mathfrak{g}_2\end{pmatrix}$, where $\mathfrak{g}_1 \in \mathcal{O}^k$ and $\mathfrak{g}_2 \in \mathcal{O}^{n-k}$, we have

$$S[\mathfrak{g}] = S_1[\mathfrak{g}_1+B\mathfrak{g}_2] + S_2[\mathfrak{g}_2] .$$

Now choose $\mathfrak{g}_2 \in \mathcal{O}^{n-k}$ satisfying $S_2[\mathfrak{g}_2] = \mu(S_2)$. Next we can find $\mathfrak{g}_1 \in \mathcal{O}^k$ according to Corollary 1.7 such that $\mathfrak{g}_1+B\mathfrak{g}_2 = (h_1,\dots,h_k)'$ and $N(h_j) \leq \frac{1}{2}$. Since S is reduced, we have $s_1 \leq \dots \leq s_k$ and $k\, s_k I \geq k \operatorname{diag} S_1 \geq S_1$ in view of Lemma 3.15. Hence

$$S_1[\mathfrak{g}_1+B\mathfrak{g}_2] \leq k\, s_k\, I[\mathfrak{g}_1+B\mathfrak{g}_2] \leq \tfrac{1}{2}k^2\, s_k \leq \tfrac{1}{2}s_{k+1}$$

according to (3). Because of $S_2[\mathfrak{g}_2] = \mu(S_2)$ the vector $\mathfrak{g}_2$ is relatively right-prime and $\mathfrak{g}$ turns out to be $(k+1)$-primitive. Now $S \in \mathcal{R}(n;\mathbb{F})$ yields

$$s_{k+1} \leq S[\mathfrak{g}] \leq \tfrac{1}{2}s_{k+1} + \mu(S_2) ,$$

$$s_{k+1} \leq 2\,\mu(S_2) .$$

By applying HERMITE's inequality to S_2 it follows that

(4) $$s_{k+1}^{n-k} \leq 2^{n-k}\, \mu(S_2)^{n-k} \leq d \det S_2 \text{ , where } d = 2^{(n-k)(n-k+1)/2} .$$

Now $S \in \mathcal{R}(n;\mathbb{F})$ implies $S_1 \in \mathcal{R}(k;\mathbb{F})$ and the induction hypothesis yields

$$s_1 \cdot \ldots \cdot s_k \leq c_k \det S_1 .$$

Regarding (2) and (4) we obtain

$$s_1 \cdot \ldots \cdot s_n \leq c\, c_k\, d \det S_1 \det S_2 = c\, c_k\, d \det S ,$$

which contradicts (1).

□

Using Corollary 3.10 each $S > 0$ possesses a unique JACOBIan representation $S = D[B]$, where $D = [d_1,\dots,d_n] > 0$ and B is an upper triangular matrix satisfying $b_{jj} = 1$.

<u>Definition.</u> Given real $\alpha > 0$ we define the <u>elementary set</u> $\mathcal{E}(n;\mathbb{F})[\alpha]$ consisting of all matrices $D[B]$, whenever $D = [d_1,\dots,d_n]$ is a diagonal matrix such that $0 < d_j < \alpha d_{j+1}$ for $1 \leq j < n$ and B is an upper triangular matrix satisfying $b_{jj} = 1$ for $1 \leq j \leq n$ as well as $N(b_{jk}) < \alpha^2$ for $1 \leq j < k \leq n$.

We have $\mathcal{E}(n;\mathbb{F})[\alpha] \subset \operatorname{Pos}(n;\mathbb{F})$ and for $\alpha > 1$ obviously

(5) $$\hat{\mathcal{E}}(n;\mathbb{F})[\alpha] = \{\hat{S} ; S \in \mathcal{E}(n;\mathbb{F})[\alpha]\} \subset \mathcal{E}(rn;\mathbb{R})[\alpha] .$$

<u>Proposition 4.9.</u> <u>There is</u> $\alpha = \alpha(n;\mathbb{F})$ <u>such that</u>
$$R(n;\mathbb{F}) \subset E(n;\mathbb{F})[\alpha] \ .$$

<u>Proof.</u> Let $S \in R(n;\mathbb{F})$ and let $S = D[B]$ denote its JACOBIan representation. Given $1 \leq k \leq n$ we have

$$s_k = d_k + \sum_{j=1}^{k-1} d_j \ N(b_{jk}) \ ,$$

hence $d_k \leq s_k$. Using the proposition above it follows that

$$(6) \qquad 1 \leq \frac{s_k}{d_k} \leq \frac{s_1 \cdot \ldots \cdot s_n}{d_1 \cdot \ldots \cdot d_n} = \frac{s_1 \cdot \ldots \cdot s_n}{\det S} \leq c_n \ .$$

Because of $s_1 \leq \ldots \leq s_n$ we obtain

$$(7) \qquad \frac{d_k}{d_l} \leq c_n \frac{s_k}{s_l} \leq c_n$$

for $1 \leq k < l \leq n$.

Now we prove by induction on j that the coefficients b_{jk} are bounded. Suppose that $|b_{lk}| := \sqrt{N(b_{lk})} < c$ for $k > l$ and $1 \leq l < j$. From $S = D[B]$ we get for $k > j$:

$$s_{jk} = d_j \ b_{jk} + \sum_{l=1}^{j-1} \bar{b}_{lj} \ d_l \ b_{lk} \ ,$$

$$b_{jk} = \frac{1}{d_j} \ s_{jk} - \sum_{l=1}^{j-1} \frac{d_l}{d_j} \ \bar{b}_{lj} \ b_{lk} \ .$$

Proposition 4.7 says that $|s_{jk}| \leq s_j$, hence $\frac{1}{d_j}|s_{jk}| \leq c_n$ in view of (6). The induction hypothesis and (7) imply

$$|b_{jk}| \leq \frac{1}{d_j} \ |s_{jk}| + \sum_{l=1}^{j-1} \frac{d_l}{d_j} \ |b_{lj}|\,|b_{lk}|$$
$$\leq c_n + (j-1) c_n \, c^2 \leq c_n (1 + nc^2) \ .$$

□

<u>Proposition 4.10.</u> <u>Let</u> $E(n;\mathbb{F})[\alpha]$ <u>be an elementary set.</u>

a) <u>There exists</u> $\beta > 0$ <u>depending only on</u> n, $\mathbb{F}$ <u>and</u> α <u>such that</u>

$$S \geq \beta \ \mathrm{diag}\ S$$

<u>for all</u> $S \in E(n;\mathbb{F})[\alpha]$.

b) *There are only finitely many* $U \in GL(n;\mathcal{O})$ *satisfying*

$$E(n;\mathbb{F})[\alpha][U] \cap E(n;\mathbb{F})[\alpha] \neq \emptyset .$$

Proof. Without restriction assume $\alpha > 1$. In view of (5) it suffices to prove part a) for $\mathbb{F} = \mathbb{R}$. In this case we refer to [21], II.1.2.

Given $U \in GL(n;\mathcal{O})$ one has $2\hat{U} \in Mat(rn;\mathbb{Z})$ and $|\det \hat{U}| = 1$. From [21], II.1.6, it follows that there are only finitely many $V \in GL(rn;\mathbb{Q})$ satisfying $2V \in Mat(rn;\mathbb{Z})$, $|\det V| = 1$ and $E(rn;\mathbb{R})[\alpha][U] \cap E(rn;\mathbb{R})[\alpha] \neq \emptyset$. Hence part b) follows in view of (5).

□

Since the coefficients of the JACOBIan representation evidently depend on $S \in Pos(n;\mathbb{F})$ in a continuous manner, we have the following

Proposition 4.11. *Given a compact subset* $C \subset Pos(n;\mathbb{F})$ *there exists* $\alpha > 0$ *such that*

$$C \subset E(n;\mathbb{F})[\alpha] .$$

Summarizing the assertions of the preceding propositions we formulate

Theorem 4.12. *Main theorem of reduction theory*
$R(n;\mathbb{F})$ *is a fundamental domain of* $Pos(n;\mathbb{F})$ *with respect to the action* $S \longmapsto S[U]$, $U \in GL(n;\mathcal{O})$, *and a convex cone. There are constants* $\alpha = \alpha(n;\mathbb{F}) > 0$ *and* $\beta = \beta(n;\mathbb{F}) > 0$ *such that together with* $c(\mathbb{R}) = \frac{1}{4}$ *and* $c(\mathbb{C}) = c(\mathbb{H}) = \frac{1}{2}$ *all* $S \in R(n;\mathbb{F})$ *satisfy:*

(i) $s_1 \leq \ldots \leq s_n$,

(ii) $N(s_{kl}) \leq c(\mathbb{F}) s_k^2$ *for* $1 \leq k < l \leq n$,

(iii) $s_1 \cdot \ldots \cdot s_n \leq \alpha \det S$,

(iv) $n \operatorname{diag} S \geq S \geq \beta \operatorname{diag} S$.

In the case $n = 2$ the domain of all reduced matrices can easily be described by inequalities. Especially one verifies

$$R(2;\mathbb{R}) = \left\{ \begin{pmatrix} s_1 & s_{12} \\ s_{12} & s_2 \end{pmatrix} \in Pos(2;\mathbb{R}) \; ; \; 0 \leq 2s_{12} \leq s_1 \leq s_2 \right\},$$

$$R(2;\mathbb{C}) = \left\{ \begin{pmatrix} s_1 & s_{12} \\ \bar{s}_{12} & s_2 \end{pmatrix} \in \mathrm{Pos}(2;\mathbb{C}) \; ; \; 2|s_{12}^{(2)}| \leq 2s_{12}^{(1)} \leq s_1 \leq s_2 \right\}.$$

$R(2;\mathbb{H})$ consists of all matrices $S = \begin{pmatrix} s_1 & s_{12} \\ \bar{s}_{12} & s_2 \end{pmatrix} \in \mathrm{Pos}(2;\mathbb{H})$ satisfying

$$s_{12} = \sum_{j=1}^{4} s_{12}^{(j)} e_j \in \mathbb{H} \; , \; 0 \leq s_{12}^{(3)} \leq s_{12}^{(2)} \, , \; |s_{12}^{(4)}| \leq s_{12}^{(2)} \quad \text{and}$$

$$2(s_{12}^{(2)} + s_{12}^{(3)} + |s_{12}^{(4)}|) \leq 2s_{12}^{(1)} \leq s_1 \leq s_2 \; .$$

In view of our applications in mind concerning modular forms we prove

<u>Lemma 4.13.</u> <u>Let</u> $S \in R(n;\mathbb{F})$. <u>The matrix</u> $S_t := \begin{pmatrix} S & 0 \\ 0 & t \end{pmatrix}$, $t \in \mathbb{R}$, <u>belongs to</u> $R(n+1;\mathbb{F})$ <u>if and only if</u> $t \geq s_n$.

<u>Proof.</u> The condition $t \geq s_n$ is necessary according to Proposition 4.7. Suppose that $t \geq s_n$ holds and a k-primitve vector $\mathfrak{g} = \begin{pmatrix} \mathfrak{h} \\ g \end{pmatrix}$, where $\mathfrak{h} \in \mathcal{O}^n$ and $g \in \mathcal{O}$, is given. If $g \neq 0$, we have $S_t[\mathfrak{g}] = S[\mathfrak{h}] + t\,N(g) \geq t \geq s_j$ for $1 \leq j \leq n$. If $g = 0$, it follows that $k \leq n$ and $S_t[\mathfrak{g}] = S[\mathfrak{h}] \geq s_k$, because S is reduced. Hence S_t belongs to $R(n+1;\mathbb{F})$. □

By induction we can determine all diagonal matrices in $R(n;\mathbb{F})$. A real diagonal matrix $D = [d_1,\ldots,d_n]$ belongs to $R(n;\mathbb{F})$ if and only if

$$0 < d_1 \leq d_2 \leq \ldots \leq d_n \; .$$

Especially, one has

$$I \in R(n;\mathbb{F})$$

for all $n \in \mathbb{N}$.

§5 Applications of reduction theory

The reduced matrices form a fundamental domain of $Pos(n;\mathbb{F})$ with respect to the action of the unimodular group. The Main theorem of reduction theory allows several applications to class invariants, i.e. functions which take the same value on every equivalence class of a matrix $S \in Pos(n;\mathbb{F})$. In §4 we have seen that $\mu(S;\mathbb{F})$ and $\det S$ are class invariants. The presentation is adapted from [37], I,§3,§5.

Let $\alpha = \alpha(n;\mathbb{F})$ and $\beta = \beta(n;\mathbb{F})$ denote the constants quoted in Theorem 4.12.

<u>Proposition 5.1.</u> a) $S \geq \beta\, \mu(S) I$ <u>for every</u> $S \in R(n;\mathbb{F})$.
b) <u>Given</u> $\gamma, \delta > 0$ <u>the set</u>

$$\{S \in R(n;\mathbb{F})\ ;\ \det S \leq \gamma\ ,\ \mu(S) \geq \delta\}$$

<u>is bounded</u>.

<u>Proof.</u> a) One has $S \geq \beta \operatorname{diag} S$ according to Theorem 4.12. Proposition 4.7 yields $\operatorname{diag} S \geq \mu(S) I$.

b) From Theorem 4.12 it follows that $s_1 \cdot \ldots \cdot s_n \leq \alpha \det S$. Hence the product is bounded. Because of $\delta \leq \mu(S) = s_1 \leq \ldots \leq s_n$ each diagonal element is bounded. Since $N(s_{kl}) \leq \frac{1}{2}s_k^2$ holds for $k \neq l$, the described set is bounded, too. □

Corollary 3.5 says $\det S \in \mathbb{Z}$ for all integral matrices $S \in Sym(n;\mathbb{F})$. Given $N \in \mathbb{N}$ it makes sense to denote the number of equivalence classes of integral matrices $S \in Pos(n;\mathbb{F})$ satisfying $\det S = N$ by $c_n(N;\mathbb{F})$, if a priori $c_n(N;\mathbb{F}) = \infty$ is admitted. $c_n(N;\mathbb{F})$ is called the <u>class number</u>.

<u>Theorem 5.2.</u> <u>The class numbers</u> $c_n(N;\mathbb{F})$, $N \geq 1$, <u>are finite</u>.

<u>Proof.</u> It suffices to estimate the number of all integral matrices $S \in R(n;\mathbb{F})$ satisfying $\det S = N$. Such an S fulfills $\mu(S) \geq 1$. Hence the assertion follows from Proposition 5.1. □

The reduction theory supplies an estimation of the class numbers, namely

$$c_n(N;\mathbb{F}) = \mathcal{O}(N^h) \quad \text{for} \quad N \longrightarrow \infty \, ,$$

where $h = n + \frac{1}{2}rn(n-1)$ and $\mathcal{O}$ denotes LANDAU's big-Oh-notation.

Given $S \in Sym(n;\mathbb{F})$ and $T \in Sym(m;\mathbb{F})$ we define

$$\Lambda(S,T;\mathbb{F}) := \{G \in Mat(n,m;\mathcal{O}) \; ; \; S[G] = T\} \, ,$$

i.e. the set of representations of T by S, and call

$$\#(S,T;\mathbb{F}) := \text{ord}\, \Lambda(S,T;\mathbb{F})$$

the _number of representations of_ T _by_ S.

Proposition 5.3. _Suppose that_ $S \in Sym(n;\mathbb{F})$ _and_ $T \in Sym(m;\mathbb{F})$. _Given_ $U \in GL(n;\mathcal{O})$ _and_ $V \in GL(m;\mathcal{O})$ _it holds that_

$$\#(S[U],T[V];\mathbb{F}) = \#(S,T;\mathbb{F}) \, .$$

_If_ S _is positive definite_ $\#(S,T;\mathbb{F})$ _becomes finite_.

Proof. The first part follows from $\Lambda(S[U],T[V];\mathbb{F}) = U^{-1}\Lambda(S,T;\mathbb{F})V$. If S is positive definite $\Lambda(S,T;\mathbb{F})$ is bounded in view of Lemma 3.14 and hence finite. □

But equivalence classes can even be separated by numbers of representations.

Theorem 5.4. _Two matrices_ $S_1, S_2 \in Pos(n;\mathbb{F})$ _are integrally equivalent if and only if_

$$\#(S_1,T;\mathbb{F}) = \#(S_2,T;\mathbb{F})$$

for all $T \in Pos(n;\mathbb{F})$.

Proof. For integrally equivalent matrices the numbers of representations coincide according to the preceding proposition. On the other hand, one uses $\#(S_1,T;\mathbb{F}) = \#(S_2,T;\mathbb{F})$ for $T = S_1$ resp. $T = S_2$ in order to see that S_1 and S_2 are integrally equivalent. □

Given $S \in Sym(n;\mathbb{F})$ one especially puts

$$Aut(S;\mathbb{F}) := \Lambda(S,S;\mathbb{F}) \, .$$

The elements of $Aut(S;\mathbb{F})$ are called the _automorphs_ of S.

<u>Proposition 5.5.</u> <u>Given</u> $S \in \mathrm{Sym}(n;\mathbb{F})$ <u>such that</u> $\det S \neq 0$ <u>then</u> $\mathrm{Aut}\, S$ <u>becomes a subgroup of</u> $GL(n;\mathcal{O})$. <u>If</u> S <u>is positive definite</u> $\mathrm{Aut}\, S$ <u>turns out to be finite</u>.

<u>Proof.</u> $U \in \mathrm{Aut}(S;\mathbb{F})$ satisfies $S[U] = S$. Now $\det S \neq 0$ implies $|\det \hat{U}| = 1$, hence $U \in GL(n;\mathcal{O})$. Obviously $\mathrm{Aut}(S;\mathbb{F})$ is a group and finite for $S > 0$ in view of Proposition 5.3. □

The Main theorem of reduction theory yields the important statement that groups of automorphs cannot grow arbitrarily.

<u>Theorem 5.6.</u> <u>There is</u> $\gamma = \gamma(n;\mathbb{F}) \in \mathbb{N}$ <u>such that</u>

$$\mathrm{ord}\, \mathrm{Aut}(S;\mathbb{F}) \leq \gamma$$

<u>for all</u> $S \in \mathrm{Pos}(n;\mathbb{F})$.

<u>Proof.</u> It suffices to prove the assertion for $S \in R(n;\mathbb{F})$, what is done by induction on n . We may put $\gamma(1;\mathbb{F}) = \mathrm{ord}\, E(\mathbb{F})$. Suppose that $n > 1$ and $U = (\mathfrak{u}_1,\ldots,\mathfrak{u}_n) \in \mathrm{Aut}(S;\mathbb{F})$. Now $S[U] = S$ yields $\mu(S) = s_1 = S[\mathfrak{u}_1] \geq \beta\, \mu(S)\, \bar{\mathfrak{u}}_1{}'\, \mathfrak{u}_1$ according to Proposition 5.1. Hence $\mathfrak{u}_1$ belongs to the finite set

$$\{\mathfrak{g} \in \mathcal{O}^n \;;\; \bar{\mathfrak{g}}'\mathfrak{g} \leq \beta^{-1}\} .$$

If $U, U_1 \in \mathrm{Aut}(S;\mathbb{F})$ possess the same first column it follows that

$$V = U_1^{-1}U = \begin{pmatrix} 1 & \bar{\mathfrak{v}}' \\ 0 & W \end{pmatrix} \in \mathrm{Aut}(S;\mathbb{F}) \text{ , where } \mathfrak{v} \in \mathcal{O}^{n-1} \text{ and } W \in GL(n-1;\mathcal{O}) .$$

Choose a decomposition

$$S = \begin{pmatrix} s_1 & \bar{\mathfrak{s}}' \\ \mathfrak{s} & T \end{pmatrix} \text{ , where } \mathfrak{s} \in \mathbb{F}^{n-1} \text{ and } T \in \mathrm{Pos}(n-1;\mathbb{F}) .$$

Now $S[V] = S$ yields

$$\mathfrak{v} = s_1^{-1}(\mathfrak{s} - \bar{W}'\mathfrak{s}) \text{ , } (T - s_1^{-1}\mathfrak{s}\bar{\mathfrak{s}}')[W] = T - s_1^{-1}\mathfrak{s}\bar{\mathfrak{s}}' .$$

Lemma 3.2 implies $T - s_1^{-1}\mathfrak{s}\bar{\mathfrak{s}}' \in \mathrm{Pos}(n-1;\mathbb{F})$. The induction hypothesis applied to $T - s_1^{-1}\mathfrak{s}\bar{\mathfrak{s}}'$ completes the proof. □

On the other hand, if G is a finite subgroup of $GL(n;\mathcal{O})$ there exists $S \in \mathrm{Pos}(n;\mathbb{F})$ such that G is a subgroup of $\mathrm{Aut}(S;\mathbb{F})$, namely

$$S := \sum_{U \in G} \bar{U}'U .$$

Choosing $t = \gamma(n;\mathbb{F})!$ therefore yields

Corollary 5.7. There exists $t = t(n;\mathbb{F}) \in \mathbb{N}$ such that $U^t = I$ holds for every $U \in GL(n;\mathcal{O})$ of finite order.

It may be very difficult to compute numbers of representations explicitly. But it is easy to check

$$\#(I^{(n)},I^{(m)};\mathbb{F}) = \frac{n!}{(n-m)!}\,(\mathrm{ord}\,E(\mathbb{F}))^m$$

for $n \geq m$. In view of Theorem 5.6 one can show that

$$\#\,\mathrm{Aut}(S;\mathbb{H}) \leq \#\,\mathrm{Aut}(S_{\mathbb{H}};\mathbb{H}) = 1920$$

for all $S \in \mathrm{Pos}(2;\mathbb{H})$, where $S_{\mathbb{H}} = \begin{pmatrix} 2 & e_1+e_2 \\ e_1-e_2 & 2 \end{pmatrix}$.

From KOECHER [36], Lemma 16, we adopt

Lemma 5.8. Let $dS = \prod_{1\leq k\leq n} ds_k \prod_{\substack{1\leq k<l\leq n\\ 1\leq j\leq r}} ds_{kl}^{(j)}$ denote the Euclidean volume element of $\mathrm{Sym}(n;\mathbb{F})$. Then it holds that

$$\nu(n;\mathbb{F}) = \int_{\substack{S\in R(n;\mathbb{F})\\ \det S\leq 1}} dS < \infty .$$

According to MAASS [43], Hilfssatz 4, we use the preceding lemma to compute some integrals.

Lemma 5.9. Let $h = n + \frac{1}{2}rn(n-1)$. Given a continuous function $\varphi : \mathbb{R}^+ \longrightarrow \mathbb{C}$ and $0<a<b$ it holds that

$$\int_{\substack{S\in R(n;\mathbb{F})\\ a\leq\det S\leq b}} \varphi(\det S)(\det S)^{-h/n}\, dS$$

$$= \frac{h}{n}\,\nu(n;\mathbb{F}) \int_a^b \varphi(s)\,\frac{ds}{s} .$$

Proof. Since $R(n;\mathbb{F})$ is a convex cone, we have

$$(1) \qquad \int_{\substack{S\in R(n;\mathbb{F})\\ \det S\leq b}} dS = \nu(n;\mathbb{F})\, b^{h/n} .$$

Given $x > a$ we define

$$\psi(x) := \int_{\substack{S \in R(n;\mathbb{F}) \\ a \leq \det S \leq x}} \varphi(\det S)(\det S)^{-h/n} \, dS .$$

Consider $t > 0$. As φ is continuous, one can find $\varepsilon \in [0;1]$ such that

$$\frac{\psi(x+t) - \psi(x)}{t} = \varphi(x+\varepsilon t)(x+\varepsilon t)^{-h/n} \frac{1}{t} \int_{\substack{S \in R(n;\mathbb{F}) \\ x \leq \det S \leq x+t}} dS$$

$$= \varphi(x+\varepsilon t)(x+\varepsilon t)^{-h/n} \frac{1}{t}\left((x+t)^{h/n} - x^{h/n}\right) \upsilon(n;\mathbb{F})$$

in view of (1). Hence it follows that

$$\psi'(x) = \frac{h}{n} \upsilon(n;\mathbb{F}) \varphi(x) x^{-1} .$$

Integration yields the assertion. □

Now we state two special cases of the lemma. We put $\varphi(x) = x^{\gamma}$, $\gamma \in \mathbb{R}$, and consider the limit $a \longrightarrow 0$ resp. $b \longrightarrow \infty$.

<u>Corollary 5.10.</u> a) <u>Given</u> $b > 0$ <u>and</u> $\gamma \in \mathbb{R}$ <u>satisfying</u> $h + n\gamma > 0$ <u>we have</u>

$$\int_{\substack{S \in R(n;\mathbb{F}) \\ \det S \leq b}} (\det S)^{\gamma} \, dS = \frac{h}{h+n\gamma} \upsilon(n;\mathbb{F}) \, b^{\gamma+h/n} .$$

b) <u>Given</u> $a > 0$ <u>and</u> $\delta \in \mathbb{R}$ <u>satisfying</u> $h + n\delta < 0$ <u>it holds that</u>

$$\int_{\substack{S \in R(n;\mathbb{F}) \\ \det S \geq a}} (\det S)^{\delta} \, dS = -\frac{h}{h+n\delta} \upsilon(n;\mathbb{F}) \, a^{\delta+h/n} .$$

Considering the trivial case $n = 1$ clearly

$$\upsilon(1;\mathbb{F}) = 1$$

is valid for $\mathbb{F} = \mathbb{R}, \mathbb{C}, \mathbb{H}$.

Chapter II   Modular group and fundamental domain

In this chapter we shall construct a fundamental domain of the half-space with respect to the action of the modular group and thus present the essential tools to investigate modular forms.

In §1 the half-space will be introduced as a suitable generalization of the upper half in the complex plane. The group of biholomorphic maps of the half-space onto itself can be described by the symplectic group, where the transposed map has to be added in some specific cases.

By means of the ordering the modular group is defined. The modular group represents a discrete subgroup of the symplectic group, which acts discontinuously on the half-space. By virtue of the Euclidean algorithm we can generalize KRONECKER's result [42] and state a simple set of generators of the modular group.

In §3 the generalization of MINKOWSKI's reduction theory is used to construct a fundamental domain of the half-space. The procedure is due to SIEGEL [54] and tends to applications to the theory of modular forms.

§4 deals with congruence subgroups of the modular group. Most of the results concerning congruence subgroups can easily be derived from the theory with respect to the full modular group.

§1 Symplectic group and half-space

The concept of half-spaces generalizes the classical case $n = 1$ of the upper half in $\mathbb{C}$. The group $SL(2;\mathbb{R}) := \{M \in Mat(2;\mathbb{R})\ ;\ \det M = 1\}$ must therefore be replaced by the symplectic group. Symplectic transformations are biholomorphic maps of the half-space onto itself. The representation is based on the description of SIEGEL's half-space in [37], Kap.III.

Let $I = I^{(n)}$ denote the identity matrix in $Mat(n;\mathbb{F})$ and put $J := \begin{pmatrix} 0 & I \\ -I & 0 \end{pmatrix} \in Mat(2n;\mathbb{F})$. Now define

$$Sp(n;\mathbb{F}) := \{M \in Mat(2n;\mathbb{F})\ ;\ J[M] = J\}.$$

<u>Lemma 1.1.</u> $Sp(n;\mathbb{F})$ <u>is a subgroup of</u> $GL(2n;\mathbb{F})$. <u>Given</u> $M = \begin{pmatrix} A & B \\ C & D \end{pmatrix}$, <u>where</u> $A,B,C,D \in Mat(n;\mathbb{F})$, <u>the following assertions are equivalent</u>:

(i) $M \in Sp(n;\mathbb{F})$.

(ii) $\bar{M}' \in Sp(n;\mathbb{F})$.

(iii) $\bar{A}'C - \bar{C}'A = \bar{B}'D - \bar{D}'B = 0\ ,\ \bar{A}'D - \bar{C}'B = I$.

(iv) $A\bar{B}' - B\bar{A}' = C\bar{D}' - D\bar{C}' = 0\ ,\ A\bar{D}' - B\bar{C}' = I$.

<u>In this case one has</u>

$$M^{-1} = \begin{pmatrix} \bar{D}' & -\bar{B}' \\ -\bar{C}' & \bar{A}' \end{pmatrix}.$$

<u>The maps</u> $Sp(n;\mathbb{F}) \longrightarrow Sp(rn;\mathbb{R})$, $M \longmapsto \hat{M}$, <u>and</u> $Sp(n;\mathbb{H}) \longrightarrow Sp(2n;\mathbb{C})$, $M \longmapsto \check{M}$, <u>are monomorphisms of the groups</u>.

<u>Proof.</u> $Sp(n;\mathbb{F})$ is contained in $GL(2n;\mathbb{F})$, because J is invertible. Clearly, $Sp(n;\mathbb{F})$ is a group. Given $M \in Sp(n;\mathbb{F})$ it follows from $J[M] = J$ that $-J = J^{-1} = (J[M])^{-1} = -J[\bar{M}'^{-1}]$, hence $\bar{M}' \in Sp(n;\mathbb{F})$. Computing $J[M]$ resp. $J[\bar{M}']$ we observe that the assertions (i) and (iii) resp. (ii) and (iv) are equivalent.

From $J[M] = J$ we obtain $M^{-1} = -J\bar{M}'J$ and hence the explicit form of M^{-1}. The last assertion holds because of $\widehat{J^{(2n)}} = J^{(2rn)}$, $\check{J}^{(2n)} = J^{(4n)}$ and Lemma I.2.1.

□

$Sp(n;\mathbb{F})$ is called the <u>symplectic group</u>. We use the term <u>fundamental relations of symplectic matrices</u> for the properties (iii) and (iv).

What about the classical case $n = 1$? The identity $Sp(1;\mathbb{R}) = SL(2;\mathbb{R})$ is well-known. Using the fundamental relations it follows that

$$(1) \qquad Sp(1;\mathbb{F}) = \{\varepsilon M \; ; \; M \in SL(2;\mathbb{R}) \; , \; \varepsilon \in \mathbb{F} \; , \; N(\varepsilon) = 1\} \; .$$

If $n \geq 1$ is arbitrary we obtain special symplectic matrices by

$$J \; , \; \begin{pmatrix} \bar{W}' & 0 \\ 0 & W^{-1} \end{pmatrix} \; , \; W \in GL(n;\mathbb{F}) \; , \; \begin{pmatrix} I & S \\ 0 & I \end{pmatrix} \; , \; S \in Sym(n;\mathbb{F}) \; .$$

Given a matrix $M = \begin{pmatrix} A & B \\ C & D \end{pmatrix} \in Mat(2n;\mathbb{F})$ we always choose a decomposition into blocks A,B,C,D belonging to $Mat(n;\mathbb{F})$. Then $(C \; D)$ is called the <u>second row</u> of M .

<u>Proposition 1.2.</u> a) <u>A matrix</u> $M = \begin{pmatrix} A & B \\ C & D \end{pmatrix} \in Mat(2n;\mathbb{F})$, <u>where</u> $C = 0$, <u>belongs to</u> $Sp(n;\mathbb{F})$ <u>if and only if there exist</u> $W \in GL(n;\mathbb{F})$ <u>and</u> $S \in Sym(n;\mathbb{F})$ <u>such that</u>

$$(2) \qquad M = \begin{pmatrix} \bar{W}' & 0 \\ 0 & W^{-1} \end{pmatrix} \begin{pmatrix} I & S \\ 0 & I \end{pmatrix} = \begin{pmatrix} I & S[W] \\ 0 & I \end{pmatrix} \begin{pmatrix} \bar{W}' & 0 \\ 0 & W^{-1} \end{pmatrix} .$$

b) <u>Two matrices</u> $M_1, M_2 \in Sp(n;\mathbb{F})$ <u>possess the same second row if and only if there is an</u> $S \in Sym(n;\mathbb{F})$ <u>satisfying</u>

$$M_1 = \begin{pmatrix} I & S \\ 0 & I \end{pmatrix} M_2 .$$

<u>Proof.</u> In both parts it suffices to show that the conditions quoted above are necessary.

a) Given $M = \begin{pmatrix} A & B \\ 0 & D \end{pmatrix} \in Sp(n;\mathbb{F})$ the fundamental relations imply $A\bar{D}' = I$ and $A\bar{B}' = B\bar{A}'$, hence $S := A^{-1}B \in Sym(n;\mathbb{F})$. Now (2) is fulfilled, if we define $W := \bar{A}'$.

b) Given $M_1, M_2 \in Sp(n;\mathbb{F})$ with the same second row, we compute

$$M_1 M_2^{-1} = \begin{pmatrix} * & * \\ 0 & I \end{pmatrix} \in Sp(n;\mathbb{F}) .$$

Hence part a) completes the proof. □

It is convenient to imbed symplectic groups of different degrees. Given $n_1, n_2 \in \mathbb{N}$ and $M_j = \begin{pmatrix} A_j & B_j \\ C_j & D_j \end{pmatrix} \in Sp(n_j;\mathbb{F})$ we define

$$M_1 \times M_2 = \begin{pmatrix} A_1 & 0 & B_1 & 0 \\ 0 & A_2 & 0 & B_2 \\ C_1 & 0 & D_1 & 0 \\ 0 & C_2 & 0 & D_2 \end{pmatrix}$$

Under obvious conditions on the numbers of rows one easily checks

Proposition 1.3. a) $M_1 \times M_2 \in Sp(n_1+n_2;\mathbb{F})$.

b) $(M_1 \times M_2)(N_1 \times N_2) = (M_1N_1) \times (M_2N_2)$.

c) $\begin{pmatrix} I & S \\ O & I \end{pmatrix} \times I = \begin{pmatrix} I & T \\ O & I \end{pmatrix}$, *where* $T = \begin{pmatrix} S & O \\ O & O \end{pmatrix}$, $S \in Sym(n;\mathbb{F})$.

d) $\begin{pmatrix} \bar{V}' & O \\ O & V^{-1} \end{pmatrix} \times I = \begin{pmatrix} \bar{W}' & O \\ O & W^{-1} \end{pmatrix}$, *where* $W = \begin{pmatrix} V & O \\ O & I \end{pmatrix}$, $V \in GL(n;\mathbb{F})$.

e) $\begin{pmatrix} O & I \\ -I & O \end{pmatrix} \times I = -MJMJM$, *where* $M = \begin{pmatrix} I & S \\ O & I \end{pmatrix}$, $S = \begin{pmatrix} I & O \\ O & O \end{pmatrix}$.

Now we can find a simple set of generators.

Lemma 1.4. *The symplectic group* $Sp(n;\mathbb{F})$ *is generated by the matrices*

$$J , \begin{pmatrix} \bar{W}' & O \\ O & W^{-1} \end{pmatrix} , W \in GL(n;\mathbb{F}) , \begin{pmatrix} I & S \\ O & I \end{pmatrix} , S \in Sym(n;\mathbb{F}) .$$

Proof. Let Δ_n denote the subgroup of $Sp(n;\mathbb{F})$ generated by the matrices quoted above. Consider $M \in Sp(n;\mathbb{F})$. If the C-block equals 0 , one has $M \in \Delta_n$ because of Proposition 1.2. Now suppose that rank $C = m > 0$. Then we can find $V, W \in GL(n;\mathbb{F})$ satisfying

$$\begin{pmatrix} \bar{W}' & O \\ O & W^{-1} \end{pmatrix} M \begin{pmatrix} \bar{V}' & O \\ O & V^{-1} \end{pmatrix} = M^* = \begin{pmatrix} A^* & B^* \\ C^* & D^* \end{pmatrix} , \quad C^* = \begin{pmatrix} I & O \\ O & O \end{pmatrix} , I = I^{(m)} .$$

Put $D^* = \begin{pmatrix} D_1 & D_2 \\ D_3 & D_4 \end{pmatrix}$, $D_1 = D_1^{(m)}$. From $C^*\overline{D^*}' \in Sym(n;\mathbb{F})$ it follows that $D_3 = 0$ and $D_1 \in Sym(m;\mathbb{F})$. Consider

$$K = M^* \begin{pmatrix} I & S \\ O & I \end{pmatrix} \left(\begin{pmatrix} O & I \\ -I & O \end{pmatrix} \times I \right) , \quad S = \begin{pmatrix} -D_1 & O \\ O & O \end{pmatrix} .$$

Now K is a symplectic matrix and its C-block equals 0 , hence $K \in \Delta_n$. We obtain $M \in \Delta_n$ from Proposition 1.3. □

As a consequence we derive the useful

Corollary 1.5. $\left\{ \begin{pmatrix} A & B \\ C & D \end{pmatrix} \in Sp(n;\mathbb{F}) \; ; \; C \in GL(n;\mathbb{F}) \right\}$ *is a dense subset of* $Sp(n;\mathbb{F})$.

<u>Proof.</u> Let $M = \begin{pmatrix} A & B \\ C & D \end{pmatrix} \in Sp(n;\mathbb{F})$ such that rank $C = m < n$. From the proof of the preceding lemma we conclude that there are $V,W \in GL(n;\mathbb{F})$ such that

$$V^{-1}C\bar{W}' = \begin{pmatrix} I & O \\ O & O \end{pmatrix}, \; I = I^{(m)}, \; V^{-1}DW^{-1} = \begin{pmatrix} D_1 & D_2 \\ O & D_4 \end{pmatrix}, \; D_1 = D_1^{(m)} .$$

Now $D_4 \in GL(n-m;\mathbb{F})$ holds because of $M \in GL(2n;\mathbb{F})$. Defining

$$S_k := \begin{pmatrix} O & O \\ O & \frac{1}{k}I \end{pmatrix} [\bar{W}'^{-1}], \; I = I^{(n-m)}, \text{ we compute}$$

$$M_k := M \begin{pmatrix} I & O \\ S_k & I \end{pmatrix} = \begin{pmatrix} * & * \\ C+DS_k & D \end{pmatrix} \in Sp(n;\mathbb{F})$$

and $\lim_{k\to\infty} M_k = M$. Furthermore,

$$C + DS_k = V \begin{pmatrix} I & * \\ O & \frac{1}{k}D_4 \end{pmatrix} \bar{W}'^{-1} \in GL(n;\mathbb{F})$$

completes the proof. □

Now we are going to define the half-space. Let

$$Mat(n;\mathbb{F})_{\mathbb{C}} := Mat(n;\mathbb{F}) \otimes_{\mathbb{R}} \mathbb{C} = \{Z = X + iY \; ; \; X,Y \in Mat(n;\mathbb{F})\}$$

denote the tensor product of the $\mathbb{R}$-algebras, where $Mat(n;\mathbb{R})_{\mathbb{C}}$ and $Mat(n;\mathbb{C})$ are identified. Given $Z = X + iY \in Mat(n;\mathbb{F})_{\mathbb{C}}$, where $X,Y \in Mat(n;\mathbb{F})$, we call $\text{Re}\, Z := X$ the <u>real part</u> of Z and $\text{Im}\, Z := Y$ the <u>imaginary part</u> of Z .

Let $Z = X + iY \in Mat(n;\mathbb{F})_{\mathbb{C}}$, then we define

$$\bar{Z}' := \bar{X}' + i\bar{Y}' ,$$

$$\tilde{Z} := \bar{X}' - i\bar{Y}' ,$$

$$\hat{Z} := \hat{X} + i\hat{Y} \in Mat(rn;\mathbb{C})$$

and in the case $\mathbb{F} = \mathbb{H}$ additionally

$$\check{Z} := \check{X} + i\check{Y} .$$

Lemma I.2.1 shows that Z is invertible in $Mat(n;\mathbb{F})_{\mathbb{C}}$ if and only if $\hat{Z}$ belongs to $GL(rn;\mathbb{C})$, i.e. $\det \hat{Z} \neq 0$.

Now we define the <u>half-space</u> $H(n;\mathbb{F})$ by

$$H(n;\mathbb{F}) := Sym(n;\mathbb{F}) + i\, Pos(n;\mathbb{F})$$
$$= \{Z = X + iY \in Mat(n;\mathbb{F})_{\mathbb{C}} \; ; \; X = \bar{X}' , \; Y = \bar{Y}' > 0\}$$

as a subset of $Mat(n;\mathbb{F})_{\mathbb{C}}$. Sometimes we call n the <u>degree</u> of $H(n;\mathbb{F})$.

The half-space $H(n;\mathbb{F})$ turns out to be an open subset of the tensor product $Sym(n;\mathbb{F}) \otimes_{\mathbb{R}} \mathbb{C}$ of the $\mathbb{R}$-vector spaces. As usual we call $H(n;\mathbb{R})$ SIEGEL'<u>s half-space</u> and $H(n;\mathbb{C})$ the <u>Hermitian half-space</u>. We shall use the term <u>half-space of quaternions</u> for $H(n;\mathbb{H})$.

In the classical case $n = 1$ we have $Sym(1;\mathbb{F}) = \mathbb{R}$, hence

(3) $$H(1;\mathbb{F}) = H(1;\mathbb{R}) \quad \text{for} \quad \mathbb{F} = \mathbb{C},\mathbb{H} ,$$

i.e. $H(1;\mathbb{F})$ turns out to be the upper half in the complex plane.

On the one hand, we have the obvious inclusions

$$H(n;\mathbb{R}) \subset H(n;\mathbb{C}) \subset H(n;\mathbb{H}) .$$

On the other hand, by means of the map $Z \longmapsto \hat{Z}$ the half-space $H(n;\mathbb{F})$ can also be imbedded into SIEGEL's half-space $H(rn;\mathbb{R})$; i.e. $H(n;\mathbb{F})$ can be regarded as an analytic submanifold of $H(rn;\mathbb{R})$. In the same way $H(n;\mathbb{H})$ is injectively mapped into $H(2n;\mathbb{C})$.

Given $Z \in H(n;\mathbb{F})$ we apply Corollary I.3.8 in order to find a matrix $W \in GL(n;\mathbb{F})$ and a uniquely determined real diagonal matrix $D = [d_1,\ldots,d_n]$ such that $d_1 \leq \ldots \leq d_n$ and

$$Z = (D + iI)[W] .$$

W is not unique, but $\bar{W}'W = \operatorname{Im} Z \in Pos(n;\mathbb{F})$. Hence

$$\det Z := \det \bar{W}'W \prod_{j=1}^{n} (d_j + i) \in \mathbb{C}$$

becomes well-defined. Corollary I.3.5 implies

$$\det \hat{Z} = (\det Z)^r \quad \text{for} \quad Z \in H(n;\mathbb{F}) ,$$

$$\det \check{Z} = (\det Z)^2 \quad \text{for} \quad Z \in H(n;\mathbb{H}) .$$

Especially $\det Z \neq 0$ holds for all $Z \in H(n;\mathbb{F})$.

<u>Definition.</u> Let $\mathcal{U}$ be an open subset of $Sym(n;\mathbb{F}) \otimes_{\mathbb{R}} \mathbb{C}$. A map $f: \mathcal{U} \longrightarrow \mathbb{C}$ is called <u>holomorphic</u> if for every $Z \in \mathcal{U}$ and every $W \in Sym(n;\mathbb{F}) \otimes_{\mathbb{R}} \mathbb{C}$ there exists an open neigborhood $\mathcal{V}$ of 0 in $\mathbb{C}$ such that $Z + \xi W \in \mathcal{U}$ for all $\xi \in \mathcal{V}$ and the map $\mathcal{V} \longrightarrow \mathbb{C}$, $\xi \longmapsto f(Z+\xi W)$, becomes holomorphic in the usual sense. A map $f: \mathcal{U} \longrightarrow \mathbb{C}^m$ is called holomorphic, if every component of f turns out to be holomorphic. A map $f: \mathcal{U} \longrightarrow \mathcal{U}$ is said to be <u>biholomorphic</u>, if f is bijective and f and f^{-1} are holomorphic.

Every $Z \in H(n;\mathbb{F})$ can be represented in the form $Z = (z_{kl})$, where $z_1,\dots,z_n \in \mathbb{C}$ and $z_{kl} = \sum_{j=1}^{r} z_{kl}^{(j)} e_j$, $z_{kl}^{(j)} \in \mathbb{C}$, $1 \leq j \leq r$, $1 \leq k < l \leq n$. Using HARTOGS' theorem $f: H(n;\mathbb{F}) \longrightarrow \mathbb{C}$ turns out to be holomorphic if f is a holomorphic function of the coordinates $z_1,\dots,z_n, z_{kl}^{(j)}$.

Holomorphic functions on the half-space of quaternions do not coincide with those holomorphic quaternionic functions introduced by FUETER.

The most important examples of holomorphic functions on half-spaces are given by

$$H(n;\mathbb{F}) \longrightarrow \mathbb{C}, \quad Z \longmapsto (\det Z)^m,$$

where $m \in \mathbb{Z}$, because $\det Z$ becomes a polynomial in the coordinates $z_1,\dots,z_n, z_{kl}^{(j)}$ in view of Theorem I.3.4.

The set $\operatorname{Bih} \mathcal{U} := \{f: \mathcal{U} \longrightarrow \mathcal{U};\ f \text{ biholomorphic}\}$ turns out to be a group. We shall determine $\operatorname{Bih} H(n;\mathbb{F})$ explicitly. But first we derive a weak version of the uniqueness theorem for holomorphic functions tailored to our purposes.

<u>Lemma 1.6.</u> <u>Let</u> $f: H(n;\mathbb{F}) \longrightarrow \mathbb{C}$ <u>be a holomorphic function satisfying</u> $f(iY) = 0$ <u>for all</u> $Y \in \operatorname{Pos}(n;\mathbb{F})$. <u>Then</u> f <u>vanishes identically</u>.

<u>Proof.</u> For reasons of simplification we write $f(z_1,\dots,z_h)$, $h = \dim_{\mathbb{R}} \operatorname{Sym}(n;\mathbb{F})$. Suppose that $f(z_1,\dots,z_m, iy_{m+1},\dots,iy_h) = 0$ for all points of this kind in $H(n;\mathbb{F})$. Choose an arbitrary point $Z = (z_1,\dots,z_{m+1}, iy_{m+2},\dots,iy_h) \in H(n;\mathbb{F})$. Since $\operatorname{Pos}(n;\mathbb{F})$ is a convex cone in $H(n;\mathbb{F})$, the set $\mathcal{V} := \{z \in \mathbb{C};\ (z_1,\dots,z_m,z,iy_{m+2},\dots,iy_h) \in H(n;\mathbb{F})\}$ becomes open and convex, too. The holomorphic function $\varphi: \mathcal{V} \longrightarrow \mathbb{C}$, $z \longmapsto f(z_1,\dots,z_m,z,iy_{m+2},\dots,iy_h)$ vanishes in all points iy satisfying $|y - y_{m+1}| < \varepsilon$ for some $\varepsilon > 0$. Applying the uniqueness theorem for holomorphic functions of one variable yields that φ vanishes identically, hence $f(Z) = 0$. An induction completes the proof. □

Now we consider the action of the symplectic group on the half-space.

<u>Theorem 1.7.</u> <u>Suppose that</u> $M = \begin{pmatrix} A & B \\ C & D \end{pmatrix}$ <u>and</u> M_1 <u>belong to</u> $Sp(n;\mathbb{F})$ <u>as well as</u> $Z = X + iY \in H(n;\mathbb{F})$. <u>Then one has</u>:

a) $M\{Z\} := CZ + D$ <u>and</u> $\widetilde{M\{Z\}} = \tilde{Z}\bar{C}' + \bar{D}'$ <u>are invertible in</u> $Mat(n;\mathbb{F})_{\mathbb{C}}$.

b) $M<Z> := (AZ + B)(CZ + D)^{-1} \in H(n;\mathbb{F})$.

c) $\operatorname{Im} M<Z> = (\widetilde{M\{Z\}})^{-1}\, Y\, (M\{Z\})^{-1}$,

$(\operatorname{Im} M<Z>)^{-1} = Y[\bar{C}'] + Y^{-1}[X\bar{C}' + \bar{D}']$.

d) $M<M_1<Z>> = (MM_1)<Z>$.

e) $(MM_1)\{Z\} = M\{M_1<Z>\}\, M_1\{Z\}$.

<u>The transformations</u> $H(n;\mathbb{F}) \longrightarrow H(n;\mathbb{F})$, $Z \longmapsto M<Z>$, <u>where</u> $M \in Sp(n;\mathbb{F})$, <u>form a subgroup of</u> $Bih\, H(n;\mathbb{F})$. <u>Given</u> $n \geq 2$ <u>two transformations</u> $Z \longmapsto M<Z>$ <u>and</u> $Z \longmapsto M_1<Z>$ <u>coincide if and only if there exists</u> $\varepsilon \in Cent\, \mathbb{F}$ <u>satisfying</u> $N(\varepsilon) = 1$ <u>and</u> $M_1 = \varepsilon M$.

<u>Proof.</u> For a proof of the case $\mathbb{F} = \mathbb{R}$ the reader is referred to [21], I,§1 . Now $\hat{M} \in Sp(rn;\mathbb{R})$ resp. $\hat{Z} \in H(rn;\mathbb{R})$ holds for $M \in Sp(n;\mathbb{F})$ resp. $Z \in H(n;\mathbb{F})$. In order to prove parts a) to e) it suffices to show that $M<Z> \in Mat(n;\mathbb{F})_{\mathbb{C}}$ is Hermitian again, because the other assertions follow from considering the map $Z \longmapsto \hat{Z}$. The fundamental relations (Lemma 1.1) yield

$$\begin{aligned} &\overline{(AZ + B)'}\,(CZ + D) \\ &= Z\bar{A}'CZ + Z\bar{A}'D + \bar{B}'CZ + \bar{B}'D \\ &= Z\bar{C}'AZ + Z\bar{C}'B + \bar{D}'AZ + \bar{D}'B \\ &= \overline{(CZ + D)'}\,(AZ + B) , \end{aligned}$$

hence $M<Z> = (AZ + B)(CZ + D)^{-1} \in Sym(n;\mathbb{F}) \otimes_{\mathbb{R}} \mathbb{C}$.

Since the components of the map $Z \longmapsto M<Z>$ are rational functions in the coefficients $z_1,\dots,z_n,z_{kl}^{(j)}$ of Z , the transformations $Z \longmapsto M<Z>$ turn out to be holomorphic functions. Because of e) $\{Z \longmapsto M<Z> ;\ M \in Sp(n;\mathbb{F})\}$ becomes a subgroup of $Bih\, H(n;\mathbb{F})$.

Given $\varepsilon \in Cent\, \mathbb{F}$, where $N(\varepsilon) = 1$, M and εM clearly induce the same transformations. On the other hand, if the transformations $Z \longmapsto M<Z>$ and $Z \longmapsto M_1<Z>$ coincide, consider $M_1^{-1}M$ and suppose that $M<Z> = Z$ for all $Z \in H(n;\mathbb{F})$. The special choice $Z = i\lambda I$, $\lambda > 0$, yields $M = \begin{pmatrix} A & 0 \\ 0 & A \end{pmatrix}$ satisfying $A\bar{A}' = I$. Hence $XA = AX$ holds for all $X \in Sym(n;\mathbb{F})$, thus $A = \varepsilon I$, $\varepsilon \in \mathbb{F}$, and $A\bar{A}' = I$ yields $N(\varepsilon) = 1$. If $n \geq 2$, we get $\varepsilon \in \text{Cent}\,\mathbb{F}$ in addition. □

It follows from the preceding proof that a transformation

$$H(1;\mathbb{F}) \longrightarrow H(1;\mathbb{F}) \ , \ Z \longmapsto M\langle Z\rangle \ ,$$

$M \in Sp(1;\mathbb{F})$, becomes the identical map if and only if $M = \varepsilon I$, where $\varepsilon \in \mathbb{F}$ and $N(\varepsilon) = 1$.

The maps $Z \longmapsto M\langle Z\rangle$, $M \in Sp(n;\mathbb{F})$, are called symplectic transformations.

Now define

$$\Pi : Mat(n;\mathbb{F})_{\mathbb{C}} \longrightarrow Mat(n;\mathbb{F})_{\mathbb{C}} \ , \ Z \longmapsto Z' \ .$$

Then Π belongs to $Bih\,H(n;\mathbb{F})$, whenever $\mathbb{F} = \mathbb{C}$, $n \geq 2$, or $\mathbb{F} = \mathbb{H}$, $n = 2$. This is clear for $\mathbb{F} = \mathbb{C}$ and for $\mathbb{F} = \mathbb{H}$, $n = 2$, we refer to Theorem I.3.11.

Given $n \geq 3$ then Π does not belong to $Bih\,H(n;\mathbb{H})$ any longer, because $Y > 0$ does not imply $Y' > 0$ in general. Consider

$$Y = \begin{pmatrix} 1 & -e_2 & -e_3 \\ e_2 & 2 & -e_4 \\ e_3 & e_4 & 3 \end{pmatrix} , \quad U = \begin{pmatrix} 1 & e_2 & e_3 \\ 0 & 1 & 0 \\ 0 & 0 & 1 \end{pmatrix} , \quad x = \begin{pmatrix} -3e_3 \\ -2e_4 \\ 1 \end{pmatrix} .$$

Now $U \in GL(3;\mathbb{H})$ and $Y[U] = [1,1,2]$ implies $Y \in Pos(3;\mathbb{H})$. But $Y'[x] = -2$ yields $Y' \notin Pos(3;\mathbb{H})$.

Theorem 1.8. $Bih\,H(n;\mathbb{F}) = \{Z \longmapsto M\langle Z\rangle \ ; \ M \in Sp(n;\mathbb{F})\}$, whenever $\mathbb{F} = \mathbb{R}$, $n \geq 1$, or $\mathbb{F} = \mathbb{H}$, $n \geq 3$.

$$Bih\,H(n;\mathbb{F}) = \{Z \longmapsto M\langle \Pi^{\varepsilon}(Z)\rangle \ ; \ M \in Sp(n;\mathbb{F}) \ , \ \varepsilon = 0,1\} ,$$

whenever $\mathbb{F} = \mathbb{C}$, $n \geq 2$, or $\mathbb{F} = \mathbb{H}$, $n = 2$.

Proof. A theorem of HIRZEBRUCH [24] says that $Bih\,H(n;\mathbb{F})$ is generated by the transformations $Z \longmapsto -Z^{-1}$, $Z \longmapsto Z + S$, $S \in Sym(n;\mathbb{F})$, and the automorphisms of the JORDAN-algebra $Sym(n;\mathbb{F})$. The automorphisms of $Sym(n;\mathbb{F})$ were determined by HERTNECK [23] and are given by $X \longmapsto X[W]$ resp. in the cases $\mathbb{F} = \mathbb{C}$, $n \geq 2$, and $\mathbb{F} = \mathbb{H}$, $n = 2$, additionally by $X \longmapsto X'[W]$, where $W \in GL(n;\mathbb{F})$. These transformations are described by symplectic matrices and Π . It remains to show that given a matrix $M \in Sp(n;\mathbb{F})$, $\mathbb{F} = \mathbb{C}$, $n \geq 2$, or $\mathbb{F} = \mathbb{H}$, $n = 2$, one can find $M^* \in Sp(n;\mathbb{F})$ satisfying $M\langle Z'\rangle' = M^*\langle Z\rangle$ for all $Z \in H(n;\mathbb{F})$. In the case $\mathbb{F} = \mathbb{C}$ we can choose $M^* = \bar{M}$. Considering $\mathbb{F} = \mathbb{H}$, $n = 2$, it suffices to demonstrate this property for the set of generators quoted in Lemma 1.4. One can choose $J^* = J$ and $\begin{pmatrix} I & S \\ 0 & I \end{pmatrix}^* = \begin{pmatrix} I & S' \\ 0 & I \end{pmatrix}$ for $S \in Sym(2;\mathbb{H})$. Given $W \in GL(2;\mathbb{H})$ one can find $W^* \in GL(2;\mathbb{H})$ satisfying $\begin{pmatrix} \bar{W}' & 0 \\ 0 & W^{-1} \end{pmatrix}^* = \begin{pmatrix} \overline{W^*}' & 0 \\ 0 & W^{*-1} \end{pmatrix}$ because of [23] . This property can also easily be checked for the generators $\begin{pmatrix} 0 & 1 \\ 1 & 0 \end{pmatrix}$, $\begin{pmatrix} 1 & 1 \\ 0 & 1 \end{pmatrix}$ and $\begin{pmatrix} a & 0 \\ 0 & 1 \end{pmatrix}$, where $0 \neq a \in \mathbb{H}$.

□

Of course the case $n = 1$ is contained in the theorem because of (3).

It is convenient to describe a bijection between the half-space $H(n;\mathbb{F})$ and the set of positive definite symplectic matrices

$$PSp(n;\mathbb{F}) := \{M \in Sp(n;\mathbb{F}) \; ; \; M = \bar{M}' > 0\} .$$

Theorem 1.9. *The map*

$$\psi : H(n;\mathbb{F}) \longrightarrow PSp(n;\mathbb{F}) \;, \; Z = X + iY \longmapsto \begin{pmatrix} Y^{-1} & 0 \\ 0 & Y \end{pmatrix} \begin{bmatrix} I & -X \\ 0 & I \end{bmatrix}$$

is bijective and

$$\psi(M\langle Z\rangle) = \psi(Z)[M^{-1}]$$

holds for all $Z \in H(n;\mathbb{F})$ *and* $M \in Sp(n;\mathbb{F})$.

Proof. The case $\mathbb{F} = \mathbb{R}$ was treated by SIEGEL [55], Lemma 6 . The remaining cases can be derived by considering $Z \longmapsto \hat{Z}$ and $M \longmapsto \hat{M}$.

□

According to BRAUN [7], III.8, we describe a volume element in $H(n;\mathbb{F})$, which is invariant under $Bih\, H(n;\mathbb{F})$.

Put $h = \dim Sym(n;\mathbb{F}) = n + \frac{1}{2} rn(n-1)$ and let

$$dX = \prod_{1 \le k \le n} dx_k \prod_{\substack{1 \le k < l \le n \\ 1 \le j \le r}} dx_{kl}^{(j)}$$ denote the Euclidean volume element in $Sym(n;\mathbb{F})$.

<u>Theorem 1.10.</u> $d\upsilon := (\det Y)^{-2h/n}\, dX\, dY$ <u>is a volume element in</u> $H(n;\mathbb{F})$, <u>which is invariant under</u> $\mathrm{Bih}\, H(n;\mathbb{F})$.

<u>Proof.</u> Denote the coefficients of $Z \in H(n;\mathbb{F})$ by $z_1,\dots,z_n, z_{kl} = \sum_{j=1}^{r} z_{kl}^{(j)} e_j$, $1 \le k < l \le n$. Then dZ (and correspondingly $d\tilde{Z}$) stands for the matrix, the elements of which are $dz_1,\dots,dz_n, \sum_{j=1}^{r} dz_{kl}^{(j)} e_j$, $1 \le k < l \le n$. First we show that $ds^2 := \tau(dZ[Y^{-1}], d\tilde{Z})$ turns out to be a differential form, which is invariant under $\mathrm{Bih}\, H(n;\mathbb{F})$. It suffices to demonstrate this property for the generators of $\mathrm{Bih}\, H(n;\mathbb{F})$ quoted in Theorem 1.8.

Considering the translations $Z \longmapsto Z + S$, $S \in \mathrm{Sym}(n;\mathbb{F})$, the assertion is obvious. Put $Z_1 = Z[U]$, $U \in GL(n;\mathbb{F})$, hence $Y_1 = Y[U]$, $Y_1^{-1} = Y^{-1}[\bar{U}'^{-1}]$, $dZ_1 = (dZ)[U]$, $d\tilde{Z}_1 = (d\tilde{Z})[U]$. Thus it follows that

$$\begin{aligned}\tau(dZ_1[Y_1^{-1}], d\tilde{Z}_1) &= \tau(dZ[Y^{-1}][\bar{U}'^{-1}], d\tilde{Z}[U]) \\ &= \tau(dZ[Y^{-1}], d\tilde{Z}) .\end{aligned}$$

Defining $Z_1 := -Z^{-1}$ Theorem 1.7 yields $Y_1^{-1} = ZY^{-1}\tilde{Z}$, $dZ_1 = Z^{-1} dZ\, Z^{-1}$, $d\tilde{Z}_1 = \tilde{Z}^{-1} d\tilde{Z}\, \tilde{Z}^{-1}$, hence

$$\begin{aligned}\tau(dZ_1[Y_1^{-1}], d\tilde{Z}_1) &= \tau(dZ[Y^{-1}][\tilde{Z}], d\tilde{Z}[\tilde{Z}^{-1}]) \\ &= \tau(dZ[Y^{-1}], d\tilde{Z}) .\end{aligned}$$

In the cases $\mathbb{F} = \mathbb{C}$, $n \ge 2$, and $\mathbb{F} = \mathbb{H}$, $n = 2$, the form ds^2 is obviously invariant under Π.

The differential form ds^2 turns out to be a positive definite quadratic form in the coordinates dx_k, dy_k, $1 \le k \le n$, $dx_{kl}^{(j)}, dy_{kl}^{(j)}$, $1 \le k < l \le n$, $1 \le j \le r$, since ds^2 is positive definite in $Z = iI$ because of

$$\tau(dZ, d\tilde{Z}) = \sum_{k=1}^{n} (dx_k^2 + dy_k^2) \quad + \quad 2 \sum_{\substack{1 \le k < l \le n \\ 1 \le j \le r}} \left({dx_{kl}^{(j)}}^2 + {dy_{kl}^{(j)}}^2 \right)$$

Let Δ denote the discriminant of the quadratic form ds^2. Thus

$$d\upsilon = \Delta^{1/2}\, dX\, dY$$

is a volume element, which is invariant under $\mathrm{Bih}\, H(n;\mathbb{F})$.

Given $Y > 0$ choose $W \in GL(n;\mathbb{F})$ satisfying $Y[W] = I$. Under the transformations $dZ \longmapsto dZ[W^{-1}]$ and $d\tilde{Z} \longmapsto d\tilde{Z}[W^{-1}]$ now ds^2 is changed into $\tau(dZ,d\tilde{Z})$. Thus Δ equals

$$\det(\bar{W}'W)^{4h/n} = (\det Y)^{-4h/n}$$

up to a constant factor. □

$d\upsilon$ is called the <u>symplectic</u> <u>volume</u> <u>element</u>. Sometimes we use the notation $d\upsilon_n$ in order to indicate the degree. In the classical case $n = 1$ one obviously has

$$d\upsilon_1 = \frac{dx\,dy}{y^2} .$$

The symplectic volume element will be utilized later, when we are going to estimate the symplectic volume of fundamental domains (cf. 3.9, 4.3, V.1.3).

§2 Modular group

In dependence on the given ordering we define the modular group as the set of integral symplectic matrices. By means of the Euclidean algorithm we can derive a result of KRONECKER [42] - often cited as Theorem of WITT [62] - where a simple set of generators of the modular group is quoted. Proofs are based on [37], III, §3.

Proposition 2.1. $\Gamma_n := \Gamma(n;\mathbb{F}) := Sp(n;\mathbb{F}) \cap Mat(2n;\mathcal{O})$ is a discrete subgroup of $Sp(n;\mathbb{F})$. Given $M \in \Gamma_n$ then $\bar{M}' \in \Gamma_n$.

Proof. Obvious, since $\det \hat{M} = 1$ holds for all $M \in \Gamma(n;\mathbb{F})$. $\square$

Γ_n is called modular group (of degree n). By modular transformations we mean the maps $Z \longmapsto M\langle Z\rangle$, $M \in \Gamma_n$. The set of modular transformations turns out to be a discontinuous subgroup of $\mathrm{Bih}\, H(n;\mathbb{F})$.

By $Sym(n;\mathcal{O}) := Sym(n;\mathbb{F}) \cap Mat(n;\mathcal{O})$ we denote the lattice of integral Hermitian matrices. Especially, the matrices

(1) $$J\,,\ \begin{pmatrix} I & S \\ O & I \end{pmatrix},\ S \in Sym(n;\mathcal{O})\,,\ \begin{pmatrix} \bar{U}' & O \\ O & U^{-1} \end{pmatrix},\ U \in GL(n;\mathcal{O})\,,$$

belong to $\Gamma(n;\mathbb{F})$. Δ_n stands for the subgroup of Γ_n generated by the matrices (1). We want to prove $\Delta_n = \Gamma_n$.

From Proposition 1.3 it follows that

(2) $$\Delta_{n_1} \times \Delta_{n_2} \subset \Delta_{n_1+n_2}\,.$$

Let $\mathfrak{e}_1,\ldots,\mathfrak{e}_{2n}$ denote the canonical basis of $\mathbb{F}^{2n}$.

Proposition 2.2. Suppose that $\mathfrak{a},\mathfrak{c} \in \mathcal{O}^n$ satisfying $\bar{\mathfrak{a}}'\mathfrak{c} = \bar{\mathfrak{c}}'\mathfrak{a}$ and $(\mathfrak{a},\mathfrak{c}) \neq 0$. If x is a gcrd$(\mathfrak{a},\mathfrak{c})$, then one can find $M \in \Delta_n$ such that

$$M\begin{pmatrix} \mathfrak{a} \\ \mathfrak{c} \end{pmatrix} = x\,\mathfrak{e}_1\,.$$

Proof. Replacing $\mathfrak{a}$ by $\mathfrak{a}x^{-1}$ and $\mathfrak{c}$ by $\mathfrak{c}x^{-1}$ one may assume that $x = 1$. Put $\mathfrak{x} = \begin{pmatrix} \mathfrak{a} \\ \mathfrak{c} \end{pmatrix}$. In view of Corollary I.2.4 there exists an

$M_1 = \begin{pmatrix} \bar{U}' & O \\ O & U^{-1} \end{pmatrix}$, $U \in GL(n;\mathcal{O})$, such that $M_1\mathfrak{x} = (a_1,\ldots,a_n,c,0,\ldots,0)'$.

Among all those matrices $M \in \Delta_n$ having the property that the last $n-1$ coefficients of Mx vanish, choose M_2 such that $N(c)$ becomes minimal. $a_1 = 0$ yields $c = 0$ by considering $J^{(2)} \times I \in \Delta_n$. If $c \neq 0$, we would have $a_1 \neq 0$. Now the fundamental relations (Lemma 1.1) together with $\bar{\mathfrak{a}}'\mathfrak{c} = \bar{\mathfrak{c}}'\mathfrak{a}$ imply $0 \neq \bar{a}_1 c \in \mathbb{Z}$. Thus there exists $s \in \mathbb{Z}$ satisfying $N(\bar{a}_1 a_1 + s\bar{a}_1 c) < N(\bar{a}_1 c)$, hence $N(a_1+sc) < N(c)$. From $\begin{pmatrix} 0 & -1 \\ 1 & s \end{pmatrix} = \begin{pmatrix} 0 & -1 \\ 1 & 0 \end{pmatrix} \begin{pmatrix} 1 & s \\ 0 & 1 \end{pmatrix} \in \Delta_1$ and (2) we get $M = \begin{pmatrix} 0 & -1 \\ 1 & s \end{pmatrix} \times I \in \Delta_n$ and $MM_2 x = (-c, a_2, \ldots, a_n, a_1+sc, 0, \ldots, 0)'$. This contradicts the choice of M_2.

Hence $c = 0$. Repeated application of Corollary I.2.4 leads to the existence of a matrix $K \in \Delta_n$ such that $Kx = (a, 0, \ldots, 0)'$. Since x is relatively right-prime, it follows from Lemma I.2.8 that a belongs to E. Hence we may assume that $a = 1$. □

Thus we have proved the crucial assertion to derive

Theorem 2.3. *The modular group* $\Gamma(n; \mathbb{F})$ *is generated by the matrices*

$$J, \quad \begin{pmatrix} I & S \\ O & I \end{pmatrix}, \quad S \in \mathrm{Sym}(n; \mathcal{O}), \quad \begin{pmatrix} \bar{U}' & O \\ O & U^{-1} \end{pmatrix}, \quad U \in GL(n; \mathcal{O}) .$$

Proof. We use induction on n. Let $n = 1$. Given $K \in \Gamma_1$ the first column of K is relatively right-prime and fulfills the condition of the preceding proposition. Hence we can find $M \in \Delta_1$ such that $MK = \begin{pmatrix} 1 & b \\ 0 & d \end{pmatrix}$. We obtain $b \in \mathbb{Z}$ and $d = 1$ from Lemma 1.1, thus $MK \in \Delta_1$ and $K \in \Delta_1$.

Given $n > 1$ and $K \in \Gamma_n$ we use an analogous argument concerning Proposition 2.2 and may assume that

$$K = \begin{pmatrix} A & B \\ C & D \end{pmatrix}, \quad A = \begin{pmatrix} 1 & \bar{\mathfrak{a}}_1' \\ 0 & A_o \end{pmatrix}, \quad B = \begin{pmatrix} b & \bar{\mathfrak{b}}_1' \\ \mathfrak{b}_2 & B_0 \end{pmatrix}, \quad C = \begin{pmatrix} 0 & \bar{\mathfrak{c}}_1' \\ 0 & C_0 \end{pmatrix},$$

$$D = \begin{pmatrix} d & \bar{\mathfrak{d}}_1' \\ \mathfrak{d}_2 & D_0 \end{pmatrix},$$

where the blocks are built up in the same way. Now consider Lemma 1.1: $\bar{A}'C = \bar{C}'A$ leads to $\mathfrak{c}_1 = 0$ and $\bar{A}_0'C_0 = \bar{C}_0'A_0$. We obtain $d = 1$, $\mathfrak{d}_1 = 0$ and $\bar{A}_0'D_0 - \bar{C}_0'B_0 = I$ from $\bar{A}'D - \bar{C}'B = I$. Finally we get $\bar{B}_0' D_0 = \bar{D}_0' B_0$ from $\bar{B}'D = \bar{D}'B$. Hence $K_0 = \begin{pmatrix} A_0 & B_0 \\ C_0 & D_0 \end{pmatrix} \in \Gamma_{n-1}$. Because of the induction hypothesis and (2) the matrix $I^{(2)} \times K_o^{-1}$ belongs to Δ_n. Now $K^* := (I^{(2)} \times K_o^{-1})MK \in \Gamma_n$ and the C-block of K^* equals 0. Using Proposition 1.2 we conclude that

$$K^* = \begin{pmatrix} \bar{W}' & O \\ O & W^{-1} \end{pmatrix} \begin{pmatrix} I & S \\ O & I \end{pmatrix},$$

where $S \in Sym(n;\mathcal{O})$ and $W \in GL(n;\mathcal{O})$. Hence $K^* \in \Delta_n$ and $K \in \Delta_n$. □

One can show [21], A.5.4, that SIEGEL's modular group $\Gamma(n;\mathbb{R})$ is generated by the matrices J and $\begin{pmatrix} I & S \\ O & I \end{pmatrix}$, $S \in Sym(n;\mathbb{Z})$. $\Gamma(n;\mathbb{C})$ however is not generated by J and $\begin{pmatrix} I & S \\ O & I \end{pmatrix}$, $S \in Sym(n;\mathcal{O}(\mathbb{C}))$, as otherwise every $M \in \Gamma(n;\mathbb{C})$ would fulfill $\det M = +1$.

Corollary 2.4. Let $\mathfrak{a},\mathfrak{c} \in \mathcal{O}^n$ and $\begin{pmatrix} \mathfrak{a} \\ \mathfrak{c} \end{pmatrix} \neq 0$. Then $\begin{pmatrix} \mathfrak{a} \\ \mathfrak{c} \end{pmatrix}$ is a column of a matrix $M \in \Gamma(n;\mathbb{F})$ if and only if $\bar{\mathfrak{a}}'\mathfrak{c} = \bar{\mathfrak{c}}'\mathfrak{a}$ holds and $(\mathfrak{a},\mathfrak{c})$ is relatively right-prime.

The proof can easily be deduced from Proposition 2.2.

Theorem 2.5. Let $M = \begin{pmatrix} A & B \\ C & D \end{pmatrix} \in \Gamma_n$, where $\operatorname{rank} C = m$. Then there exist $U,V \in GL(n;\mathcal{O})$, $S = \begin{pmatrix} O^{(m)} & * \\ * & * \end{pmatrix} \in Sym(n;\mathcal{O})$ and $M_1 = \begin{pmatrix} A_1 & B_1 \\ C_1 & D_1 \end{pmatrix} \in \Gamma_m$ such that $\operatorname{rank} C_1 = m$ and

$$\begin{pmatrix} I & S \\ O & I \end{pmatrix} \begin{pmatrix} \bar{U}' & O \\ O & U^{-1} \end{pmatrix} M \begin{pmatrix} \bar{V}' & O \\ O & V^{-1} \end{pmatrix} = M_1 \times I .$$

Proof. In the case $m = 0$ consider Proposition 1.2. Let $0 < m < n$. In view of Theorem I.2.3 there exist $U,V \in GL(n;\mathcal{O})$ such that

$$U^{-1}C\bar{V}' = \begin{pmatrix} C_1 & O \\ 0 & 0 \end{pmatrix}, \quad C_1 = C_1^{(m)}, \quad \det \hat{C}_1 \neq 0 .$$

Put $M_o = \begin{pmatrix} \bar{U}' & O \\ O & U^{-1} \end{pmatrix} M \begin{pmatrix} \bar{V}' & O \\ O & V^{-1} \end{pmatrix} = \begin{pmatrix} A_o & B_o \\ C_o & D_o \end{pmatrix}$ and choose a decomposition into blocks $A_o = \begin{pmatrix} A_1 & A_2 \\ A_3 & A_4 \end{pmatrix}$, $A_1 = A_1^{(m)}$ etc. We apply Lemma 1.1 to M_o. Since $C_o\bar{D}_o'$ is Hermitian and C_1 is invertible, we obtain $D_3 = 0$.

Now $\bar{A}_o'C_o \in \mathrm{Sym}(n;\mathcal{O})$ leads to $\bar{A}_1'C_1 \in \mathrm{Sym}(m;\mathcal{O})$ and $A_2 = 0$ and $\bar{B}_o'D_o = \bar{D}_o'B_o$ implies $\bar{B}_1'D_1 \in \mathrm{Sym}(m;\mathcal{O})$ and $B_4 = \bar{B}_4'$. Finally we obtain $\bar{A}_1'D_1 - \bar{C}_1'B_1 = I$ and $A_4 = I$ from $\bar{A}_o'D_o - \bar{C}_o'B_o = I$. Hence $M_1 = \begin{pmatrix} A_1 & B_1 \\ C_1 & D_1 \end{pmatrix}$ belongs to Γ_m. Thus we have

$$K := M_o(M_1 \times I)^{-1} = \begin{pmatrix} I & 0 & 0 & B_2 \\ A_5 & I & B_5 & B_4 \\ 0 & 0 & I & 0 \\ 0 & 0 & 0 & I \end{pmatrix} \in \Gamma_n .$$

The fundamental relations applied to K yield $A_5 = 0$ and

$S = \begin{pmatrix} 0 & B_2 \\ B_5 & B_4 \end{pmatrix} \in \mathrm{Sym}(n;\mathcal{O})$. □

What about the modular groups in the classical case $n = 1$? The matrices εI , $\varepsilon \in E$, commute with the matrices in $SL(2;\mathbb{Z})$. Since $\Gamma(1;\mathbb{F})$ is generated by $\begin{pmatrix} 0 & 1 \\ -1 & 0 \end{pmatrix}$, $\begin{pmatrix} 1 & 1 \\ 0 & 1 \end{pmatrix}$ and εI , $\varepsilon \in E$, we state the result as

<u>Lemma 2.6.</u> $\Gamma(1;\mathbb{F}) = \{\varepsilon M \; ; \; \varepsilon \in E(\mathbb{F}) \; , \; M \in SL(2;\mathbb{Z})\}$ <u>for</u> $\mathbb{F} = \mathbb{C},\mathbb{H}$.

Thus the groups of modular transformations $Z \longmapsto M\langle Z\rangle$, $M \in \Gamma(1;\mathbb{F})$, coincide for $\mathbb{F} = \mathbb{R},\mathbb{C},\mathbb{H}$, i.e. the factor groups

$$\Gamma(1;\mathbb{F}) \Big/ \{\varepsilon I \; ; \; \varepsilon \in E(\mathbb{F})\}$$

are isomorphic.

§3 The fundamental domain $\mathcal{F}(n;\mathbb{F})$

Following SIEGEL [54] a fundamental domain of the half-space with respect to the action of the modular group is constructed. This special fundamental domain possesses some important properties, which are used in the theory of modular forms.

The construction of SIEGEL's fundamental domain is adopted from CHRISTIAN [13], Kap. IV, MAASS [45], §12, and KOECHER [37], III, §4.

In dependence on the given ordering we define the fundamental parallelotope or unit cube in $\mathrm{Sym}(n;\mathbb{F})$. Let $\mathcal{C}(n;\mathbb{F})$ denote the set of matrices $X = (x_{kl}) \in \mathrm{Sym}(n;\mathbb{F})$ obeying the conditions $-\frac{1}{2} \leq x_k \leq \frac{1}{2}$ for $1 \leq k \leq n$ and

$$x_{kl} = \sum_{j=1}^{r} x_{kl}^{(j)} e_j \quad , \quad -\tfrac{1}{2} \leq x_{kl}^{(j)} \leq \tfrac{1}{2} \quad , \quad 1 \leq k < l \leq n \quad , \quad 1 \leq j \leq r \quad ,$$

in the cases $\mathbb{F} = \mathbb{R}, \mathbb{C}$ resp.

$$x_{kl} = \sum_{j=0}^{3} x_{kl}^{(j)} e_j \quad , \quad -\tfrac{1}{2} \leq x_{kl}^{(j)} \leq \tfrac{1}{2} \quad , \quad 1 \leq k < l \leq n \quad , \quad 0 \leq j \leq 3 \quad ,$$

in the case $\mathbb{F} = \mathbb{H}$. Writing $\mathrm{Sym}(n;\mathcal{O}) := \mathrm{Sym}(n;\mathbb{F}) \cap \mathrm{Mat}(n;\mathcal{O})$ for the lattice of integral Hermitian matrices then $\mathcal{C}(n;\mathbb{F})$ is a fundamental parallelotope with respect to this lattice, i.e. $\mathcal{C}(n;\mathbb{F})$ turns out to be a fundamental domain of $\mathrm{Sym}(n;\mathbb{F})$ with respect to the action $X \longmapsto X + S$, $S \in \mathrm{Sym}(n;\mathcal{O})$.

Let $\mathcal{R}(n;\mathbb{F})$ again stand for the domain of reduced matrices in $\mathrm{Pos}(n;\mathbb{F})$. We define the set $\mathcal{F}(n;\mathbb{F})$ to consist of all matrices $Z = X + iY \in H(n;\mathbb{F})$ satisfying

(i) $X \in \mathcal{C}(n;\mathbb{F})$,

(ii) $Y \in \mathcal{R}(n;\mathbb{F})$,

(iii) $|\det \widehat{M\{Z\}}| \geq 1$ for every $M \in \Gamma(n;\mathbb{F})$.

Clearly, $\mathcal{F}(n;\mathbb{F})$ is a closed subset of the half-space $H(n;\mathbb{F})$. We are going to prove in several steps that $\mathcal{F}(n;\mathbb{F})$ is a fundamental domain of $H(n;\mathbb{F})$ with respect to the action of $\Gamma(n;\mathbb{F})$.

Proposition 3.1. There is a $\rho = \rho(n;\mathbb{F})$ such that

$$\operatorname{Im} Z \geq \rho I$$

holds for all $Z \in F(n;\mathbb{F})$. Especially, $F(n;\mathbb{F})$ is a closed subset of $\operatorname{Sym}(n;\mathbb{F}) \otimes_{\mathbb{R}} \mathbb{C}$.

Proof. Let $Z = X + iY \in F(n;\mathbb{F})$ and consider $M := J^{(2)} \times I^{(2n-2)} \in \Gamma(n;\mathbb{F})$. Then one has $1 \leq |\det \widehat{M\{Z\}}|^2 = |z_1|^{2r} = (x_1^2 + y_1^2)^r$. From $X \in C(n;\mathbb{F})$ it follows that $x_1^2 \leq \frac{1}{4}$, hence $y_1 \geq \frac{1}{2}\sqrt{3}$. Propositions I.4.7 and I.5.1 yield

$$Y \geq \tfrac{1}{2}\sqrt{3}\,\beta\, I .$$

□

Lemma 3.2. a) $iI \in F(n;\mathbb{F})$.
b) Given $Z = X + iY \in F(n;\mathbb{F})$ then $Z_\lambda := X + i\lambda Y \in F(n;\mathbb{F})$ holds for all $\lambda \geq 1$.
c) $F(n;\mathbb{F})$ is connected.

Proof. Let $Z = X + iY \in F(n;\mathbb{F})$. Since $R(n;\mathbb{F})$ is a cone, we have $\lambda Y \in R(n;\mathbb{F})$ for $\lambda \geq 1$. Given $M = \begin{pmatrix} * & * \\ C & D \end{pmatrix} \in \Gamma(n;\mathbb{F})$, where rank $C = m > 0$, we apply Theorem 2.5 in order to find $U, V \in GL(n;\mathcal{O})$ and $M_1 = \begin{pmatrix} * & * \\ C_1 & D_1 \end{pmatrix} \in \Gamma(m;\mathbb{F})$ such that $\det \hat{C}_1 \neq 0$,

$$C = U \begin{pmatrix} C_1 & 0 \\ 0 & 0 \end{pmatrix} \bar{V}' , \quad D = U \begin{pmatrix} D_1 & 0 \\ 0 & I \end{pmatrix} V^{-1} .$$

Let Q denote the matrix consisting of the first m columns of V, hence $V = (Q,*)$, $Q = Q^{(n,m)}$. Then one has

$$|\det \widehat{M\{Z\}}| = \left|\det \widehat{\left(\begin{pmatrix} C_1 & 0 \\ 0 & 0 \end{pmatrix} Z[V] + \begin{pmatrix} D_1 & 0 \\ 0 & I \end{pmatrix}\right)}\right|$$

$$= |\det \widehat{(C_1 Z[Q] + D_1)}|$$

$$= |\det \hat{C}_1|\ |\det (Z[Q] + C_1^{-1} D_1)|^r ,$$

since Lemma 1.1 yields $C_1^{-1} D_1 \in \operatorname{Sym}(m;\mathbb{F})$. Using Corollary I.3.8 there exist $W \in GL(m;\mathbb{F})$ and a real diagonal matrix $F = [f_1, \ldots, f_m]$ satisfying $Z[Q] + C_1^{-1} D_1 = (F + iI)[W]$. Considering $\lambda \geq 1$ we observe that

$$Z_\lambda [Q] + C_1^{-1} D_1 = (F + i\lambda I)[W] ,$$

hence

$$|\det \widehat{M\{Z_\lambda\}}|^2 = (\det \hat{C}_1)^2 \, (\det \bar{W}'W)^r \prod_{j=1}^{m} (f_j^2 + \lambda^2)^r$$

$$\geq (\det \hat{C}_1)^2 \, (\det \bar{W}'W)^r \prod_{j=1}^{m} (f_j^2 + 1)^r$$

$$= |\det \widehat{M\{Z\}}|^2 .$$

Thus we have $Z_\lambda \in F(n;\mathbb{F})$ for $\lambda \geq 1$.

In the case iI we compute that

$$|\det \widehat{M\{iI\}}| = |\det \hat{C}_1| \; |\det(C_1^{-1}D_1 + i\bar{Q}'Q)|^r .$$

Corollary I.2.6 leads to $|\det \hat{C}_1| \geq 1$. Now Corollary I.3.8 implies that $|\det Z| \geq \det Y$ for $Z \in F(n;\mathbb{F})$, hence

$$|\det \widehat{M\{iI\}}| \geq (\det \bar{Q}'Q)^r \geq 1$$

because of $\bar{Q}'Q \in \mathrm{Pos}(m;\mathbb{F}) \cap \mathrm{Mat}(m;\mathcal{O})$. Since I belongs to $R(n;\mathbb{F})$ in view of Lemma I.4.13, we have $iI \in F(n;\mathbb{F})$.

Let $\rho = \rho(n;\mathbb{F}) \leq 1$ be chosen as in Proposition 3.1. Given $Z = X + iY \in F(n;\mathbb{F})$ then $Z_\lambda = X + i\lambda Y$ belongs to $F(n;\mathbb{F})$ for $1 \leq \lambda \leq \rho^{-1}$. Therefore it suffices to show that $W_\lambda := \lambda X + i\lambda\rho^{-1}Y + i(1-\lambda)I$ is an element of $F(n;\mathbb{F})$ for $0 \leq \lambda \leq 1$. Clearly $\lambda X \in C(n;\mathbb{F})$ and $\lambda\rho^{-1}Y + (1-\lambda)I \in R(n;\mathbb{F})$ hold , because $Z \in F(n;\mathbb{F})$ and $R(n;\mathbb{F})$ is a convex cone. Considering $M \in \Gamma(n;\mathbb{F})$ as above we compute that

$$|\det \widehat{M\{W_\lambda\}}| = |\det \hat{C}_1| \; |\det(W_\lambda[Q] + C_1^{-1}D_1)|^r$$

$$\geq (\det(\lambda\rho^{-1}Y + (1-\lambda)I)[Q])^r .$$

The preceding proposition yields $\rho^{-1}Y \geq I$, hence $\lambda\rho^{-1}Y + (1-\lambda)I \geq I$ and

$$|\det \widehat{M\{W_\lambda\}}| \geq (\det \bar{Q}'Q)^r \geq 1 .$$

Thus we have $W_\lambda \in F(n;\mathbb{F})$ for $0 \leq \lambda \leq 1$.

□

Next we want to map an arbitrary point of the half-space into $F(n;\mathbb{F})$ by a modular transformation. Therefore we first map the point onto an equivalent point of "maximal height", i.e. a point, the imaginary part of which possesses the maximal determinant among all equivalent points.

Proposition 3.3. Given $Z \in H(n;\mathbb{F})$ there is an $M_o \in \Gamma(n;\mathbb{F})$ such that

$$\det(\mathrm{Im}\, M\langle Z\rangle) \leq \det(\mathrm{Im}\, M_o\langle Z\rangle)$$

holds for all $M \in \Gamma(n;\mathbb{F})$.

Proof. Considering Theorem 1.7 we may replace M by $\begin{pmatrix} I & S \\ O & I \end{pmatrix}\begin{pmatrix} \bar{U}' & O \\ O & U^{-1} \end{pmatrix} M$, where $S \in \mathrm{Sym}(n;\mathcal{O})$ and $U \in GL(n;\mathcal{O})$, without changing $\det \mathrm{Im}\, M\langle Z\rangle$. Choosing suitable matrices U and S it suffices to show that there are only finitely many $M \in \Gamma(n;\mathbb{F})$ satisfying

(i) $\det(\mathrm{Im}\, M\langle Z\rangle) \geq \det Y$,

(ii) $(\mathrm{Im}\, M\langle Z\rangle)^{-1} \in \mathcal{R}(n;\mathbb{F})$,

(iii) $\mathrm{Re}\, M\langle Z\rangle \in \mathcal{C}(n;\mathbb{F})$.

Given $M \in \Gamma(n;\mathbb{F})$ obeying these conditions we define $Y_M := (\mathrm{Im}\, M\langle Z\rangle)^{-1}$. Using Theorem 1.7 and standard notations yields

$$Y_M = Y^{-1}[X\bar{C}' + \bar{D}'] + Y[\bar{C}'] .$$

Now choose $x \in \mathcal{O}^n - \{0\}$ such that $\mu(Y_M) = Y_M[x]$. Putting $\mathfrak{c} := \bar{C}'x$ and $\mathfrak{d} := \bar{D}'x$ we have

$$\mu(Y_M) = Y^{-1}[X\mathfrak{c} + \mathfrak{d}] + Y[\mathfrak{c}] .$$

$\mathfrak{c} \neq 0$ leads to $\mu(Y_M) \geq Y[\mathfrak{c}] \geq \mu(Y)$. $\bar{M}' \in GL(2n;\mathcal{O})$ and $x \neq 0$ imply $\bar{M}'\begin{pmatrix} O \\ x \end{pmatrix} = \begin{pmatrix} \mathfrak{c} \\ \mathfrak{d} \end{pmatrix} \neq 0$. Hence $\mathfrak{c} = 0$ yields $\mathfrak{d} \neq 0$ and $\mu(Y_M) = Y^{-1}[\mathfrak{d}] \geq \mu(Y^{-1})$. Because of $\det Y_M \leq (\det Y)^{-1}$, $\mu(Y_M) \geq \min\{\mu(Y),\mu(Y^{-1})\}$ and $Y_M \in \mathcal{R}(n;\mathbb{F})$ the set consisting of all these Y_M is bounded in view of Proposition I.5.1. Hence there exists $\alpha > 0$ satisfying $\alpha I \geq Y_M$ for all M of the described kind. Therefore we have $\alpha I \geq Y[\bar{C}']$ and $\alpha I \geq Y^{-1}[X\bar{C}' + \bar{D}']$. Lemma I.3.14 yields that the sets consisting of all these C resp. these $CX + D$ are bounded. Hence there are only finitely many second rows of the matrices M . Two matrices in $\Gamma(n;\mathbb{F})$ having the same second row differ from each other by a left factor $\begin{pmatrix} I & S \\ O & I \end{pmatrix}$, where $S \in \mathrm{Sym}(n;\mathcal{O})$. In view of (iii) there are only finitely many matrices M satisfying the properties (i) to (iii). □

Thus we have the crucial assertion in order to prove

Lemma 3.4. Given $Z \in H(n;\mathbb{F})$ there exists an $M \in \Gamma(n;\mathbb{F})$ such that

$M\langle Z\rangle \in F(n;\mathbb{F})$.

Proof. According to the proposition above choose an $M_0 \in \Gamma(n;\mathbb{F})$ satisfying $\det(\operatorname{Im} M\langle Z\rangle) \leq \det(\operatorname{Im} M_o\langle Z\rangle)$ for all $M \in \Gamma(n;\mathbb{F})$, where M_0 may be replaced by $\begin{pmatrix} I & S \\ O & I \end{pmatrix} \begin{pmatrix} \bar{U}' & O \\ O & U^{-1} \end{pmatrix} M_0$, $S \in \operatorname{Sym}(n;\mathcal{O})$, $U \in GL(n;\mathcal{O})$. Defining $Z_0 := M_o\langle Z\rangle$ one has $\det \operatorname{Im} M\langle Z_o\rangle \leq \det \operatorname{Im} Z_0$, hence $|\det \widehat{M\{Z_0\}}| \geq 1$ for all $M \in \Gamma(n;\mathbb{F})$ in view of Theorem 1.7. Finally choose $U \in GL(n;\mathcal{O})$ such that $Y_0[U] \in R(n;\mathbb{F})$ and determine $S \in \operatorname{Sym}(n;\mathcal{O})$ in order to obtain $X_0[U] + S \in C(n;\mathbb{F})$. □

From the preceding lemma it follows that

$$H(n;\mathbb{F}) = \bigcup_M M\langle F(n;\mathbb{F})\rangle ,$$

where the union is taken over all $M \in \Gamma(n;\mathbb{F})$. Since $F(n;\mathbb{F})$ is closed, we conclude that $F(n;\mathbb{F})$ contains interior points.

According to Theorem 1.7 a modular transformation $Z \longmapsto M\langle Z\rangle$, $M \in \Gamma(n;\mathbb{F})$, is the identical map if and only if $M = \varepsilon I$, where $\varepsilon \in E(\mathbb{F})$ and for $n \geq 2$ in addition $\varepsilon \in \operatorname{Cent} \mathbb{F}$.

Clearly the boundary of $F(n;\mathbb{F})$ consists of all matrices $Z = X + iY \in F(n;\mathbb{F})$, where X belongs to the boundary of $C(n;\mathbb{F})$ or Y is contained in the boundary of $R(n;\mathbb{F})$ or there exists an $M = \begin{pmatrix} * & * \\ C & D \end{pmatrix} \in \Gamma(n;\mathbb{F})$ satisfying $C \neq 0$ and $|\det \widehat{M\{Z\}}| = 1$.

Proposition 3.5. If Z_0, Z_1 belong to $F(n;\mathbb{F})$ and the modular transformation $Z \longmapsto M\langle Z\rangle$, $M \in \Gamma(n;\mathbb{F})$, is not the identical map such that $Z_1 = M\langle Z_o\rangle$, then Z_0 and Z_1 are contained in the boundary of $F(n;\mathbb{F})$.

Proof. From $Z_1 = M\langle Z_o\rangle$ and Theorem 1.7 it follows that $(M\{Z_0\})^{-1} = M^{-1}\{Z_1\}$. Since Z_0, Z_1 belong to $F(n;\mathbb{F})$, we have $|\det \widehat{M\{Z_0\}}| = |\det \widehat{M^{-1}\{Z_1\}}| = 1$. Thus we may assume that $C = 0$.

Following Proposition 1.2 we obtain a representation

$$M = \begin{pmatrix} I & S \\ O & I \end{pmatrix} \begin{pmatrix} U' & O \\ O & U^{-1} \end{pmatrix},$$

where $S \in Sym(n;\mathcal{O})$ and $U \in GL(n;\mathcal{O})$. Now $Z_1 = Z_0[U] + S$ yields $Y_1 = Y_0[U]$. Hence we may assume that $X \longmapsto X[U]$ is the identical map. Then we get $S \neq 0$ and X_0, X_1 belong to the boundary of $C(n;\mathbb{F})$.

□

In connection with the elementary set $E(n;\mathbb{F})[\alpha]$ defined in chapter I, §4, we choose the notation of SIEGEL's elementary set for

$$S(n;\mathbb{F})[\alpha] := \{Z \in H(n;\mathbb{F})\ ;\ N(x_{kl}) < \alpha^2\ ,\ Y \in E(n;\mathbb{F})[\alpha]\ ,\ 1 < \alpha y_1\},$$

where $\alpha > 0$. Given $\alpha > 1$ we have again

$$\hat{S}(n;\mathbb{F})[\alpha] = \{\hat{Z}\ ;\ Z \in S(n;\mathbb{F})[\alpha]\} \subset S(rn;\mathbb{R})[\alpha].$$

Proposition 3.6. a) *There is an* $\alpha = \alpha(n;\mathbb{F}) > 0$ *satisfying*

$$F(n;\mathbb{F}) \subset S(n;\mathbb{F})[\alpha].$$

b) *Given a compact subset* C *of* $H(n;\mathbb{F})$ *there exists a* $\beta > 0$ *such that*

$$C \subset S(n;\mathbb{F})[\beta].$$

Proof. a) According to Propostion I.4.9 there exists an $\alpha > 0$ such that $R(n;\mathbb{F}) \subset E(n;\mathbb{F})[\alpha]$. From the proof Proposition 3.1 it follows that $y_1 \geq \frac{1}{2}\sqrt{3}$ and $X \in C(n;\mathbb{F})$ yields $N(x_{kl}) < 2$. Hence it suffices to choose $\alpha \geq \sqrt{2}$.

b) If C is compact in $H(n;\mathbb{F})$, then the sets $\{Re\ Z\ ;\ Z \in C\}$ resp. $\{Im\ Z\ ;\ Z \in C\}$ turn out to be compact in $Sym(n;\mathbb{F})$ resp. $Pos(n;\mathbb{F})$, too. Proposition I.4.11 yields the assertion.

□

Now we use the map

$$\psi : H(n;\mathbb{F}) \longrightarrow PSp(n;\mathbb{F})\ ,\ Z = X + iY \longmapsto \begin{pmatrix} Y^{-1} & O \\ O & Y \end{pmatrix} \begin{bmatrix} I & -X \\ O & I \end{bmatrix}$$

examined in Theorem 1.9. In addition we define

$$V^* := \begin{pmatrix} O & & 1 \\ & \cdot^{\cdot^{\cdot}} & \\ 1 & & O \end{pmatrix} \in GL(n;\mathcal{O}) \quad \text{and} \quad W^* = \begin{pmatrix} V^* & O \\ O & I \end{pmatrix} \in GL(2n;\mathcal{O}).$$

<u>Proposition 3.7.</u> <u>Given</u> $\alpha > 0$ <u>there exists</u> $\beta > 0$ <u>such that</u>

$$\Psi(S(n;\mathbb{F})[\alpha])[W^*] \subset E(2n;\mathbb{F})[\beta] .$$

<u>Proof.</u> Given $Z = X + iY$ let $Y = D[B]$ denote the JACOBIan representation. Then one has $Y^{-1} = D^{-1}[\bar{B}'^{-1}]$, $Y^{-1}[V^*] = F[G]$, where $F = D^{-1}[V^*]$ and $G = V^*\bar{B}'^{-1}V^*$. But G turns out to be an upper triangular matrix having 1's on the diagonal. Hence $F[G]$ is the JACOBIan representation of $Y^{-1}[V^*]$. Thus the JACOBIan representation of $\Psi(Z)[W^*]$ equals

$$\begin{pmatrix} Y^{-1} & 0 \\ 0 & Y \end{pmatrix} \begin{bmatrix} I & -X \\ 0 & I \end{bmatrix} \begin{bmatrix} V^* & 0 \\ 0 & I \end{bmatrix} = \begin{pmatrix} F & 0 \\ 0 & D \end{pmatrix} \begin{bmatrix} G & -V^*\bar{B}'^{-1}X \\ 0 & B \end{bmatrix} .$$

Considering $Z \in S(n;\mathbb{F})[\alpha]$ the set of the B's appearing in the decomposition is bounded. Because of $\det \hat{B} = 1$ the sets consisting of all $\bar{B}'^{-1}$ resp. $\bar{B}'^{-1}X$ turns out to be bounded, too. Hence the norms of the coefficients are bounded by some β^2 , $\beta > 0$, depending only on α . From $d_j < \alpha d_{j+1}$ it follows that $d_{j+1}^{-1} < \alpha\, d_j^{-1}$. Choosing $\beta \geq \alpha^2$ leads to $d_1^{-1} < \alpha \leq \beta\alpha^{-1} < \beta d_1$. From $F = [d_n^{-1},\dots,d_1^{-1}]$ we obtain $\Psi(Z)[W^*] \in E(2n;\mathbb{F})[\beta]$. □

Observing $\Gamma(n;\mathbb{F}) \subset GL(2n;\mathcal{O})$ the preceding proposition, Theorem 1.9 and Proposition I.4.10 imply

<u>Corollary 3.8.</u> <u>Given</u> $\alpha > 0$ <u>there are only finitely many</u> $M \in \Gamma(n;\mathbb{F})$ <u>satisfying</u>

$$M\langle S(n;\mathbb{F})[\alpha]\rangle \cap S(n;\mathbb{F})[\alpha] \neq \emptyset .$$

Next we estimate the symplectic volume of $F(n;\mathbb{F})$, where the symplectic volume element dv was defined in Theorem 1.10.

<u>Proposition 3.9.</u> $\operatorname{vol}(\Gamma(n;\mathbb{F})) := \int_{F(n;\mathbb{F})} dv < \infty$.

<u>Proof.</u> Using Proposition 3.1 we get an upper bound of the symplectic volume by

$$\int (\det Y)^{-2h/n}\, dX\, dY ,$$

where the integral is taken over $X \in C(n;\mathbb{F})$ and $Y \in R(n;\mathbb{F})$ such that $\det Y \geq c$ for some $c > 0$. Now $C(n;\mathbb{F})$ is compact and the integral

$$\int\limits_{\substack{Y \in \mathcal{R}(n;\mathbb{F}) \\ \det Y \geq c}} (\det Y)^{-2h/n} \, dY$$

converges according to Corollary I.5.10. □

Since dv is invariant under all symplectic transformations, the definition of $\mathrm{vol}(\Gamma(n;\mathbb{F}))$ does not depend on the choice of the special fundamental domain $\mathcal{F}(n;\mathbb{F})$, but only on the group $\Gamma(n;\mathbb{F})$.

We summarize the preceding assertions in order to formulate

Theorem 3.10. $\mathcal{F}(n;\mathbb{F})$ is a fundamental domain of the half-space $H(n;\mathbb{F})$ with respect to the action of the modular group $\Gamma(n;\mathbb{F})$ and $\mathcal{F}(n;\mathbb{F})$ possesses finite symplectic volume. $\mathcal{F}(n;\mathbb{F})$ is connected and closed in $\mathrm{Sym}(n;\mathbb{F}) \otimes_{\mathbb{R}} \mathbb{C}$.

The following assertion is of essential importance to the theory of modular forms.

Lemma 3.11. Given $Z \in \mathcal{F}(n;\mathbb{F})$ there exists $\lambda_o > 0$ such that

$$Z_\lambda := [Z, i\lambda] \in \mathcal{F}(n+1;\mathbb{F})$$

for $\lambda \geq \lambda_o$.

Proof. Put $Z_\lambda = X_\lambda + iY_\lambda$, then $X_\lambda \in \mathcal{C}(n+1;\mathbb{F})$ and $Y_\lambda \in \mathcal{R}(n+1;\mathbb{F})$ for $\lambda \geq y_n$ in view of Lemma I.4.13. Suppose that

$$M = \begin{pmatrix} * & * \\ C & D \end{pmatrix} \in \Gamma(n+1;\mathbb{F}), \text{ where } \operatorname{rank} C = m > 0 .$$

By analogy to the proof of Lemma 3.2 choose U and $V = (Q,*)$ in $GL(n+1;\mathcal{O})$ as well as $\begin{pmatrix} * & * \\ C_1 & D_1 \end{pmatrix} \in \Gamma(m;\mathbb{F})$ such that $Q = Q^{(n+1,m)}$ and

$$C = U \begin{pmatrix} C_1 & O \\ 0 & 0 \end{pmatrix} \bar{V}', \quad D = U \begin{pmatrix} D_1 & O \\ 0 & I \end{pmatrix} V^{-1}, \quad \operatorname{rank} C_1 = m .$$

Similarly we obtain $|\det \widehat{M\{Z_\lambda\}}| \geq (\det Y_\lambda[Q])^r$. We may replace Q by QW, $W \in GL(m;\mathcal{O})$. Therefore we can assume that $Y_\lambda[Q] \in \mathcal{R}(m;\mathbb{F})$ holds. Put $Q = (\mathfrak{q}_1,\ldots,\mathfrak{q}_m) = \begin{pmatrix} P \\ \bar{\mathfrak{q}}' \end{pmatrix}$, where $P = (\mathfrak{p}_1,\ldots,\mathfrak{p}_m)$, $\bar{\mathfrak{q}}' = (q_1,\ldots q_m)$, $\mathfrak{q}_j \in \mathcal{O}^{n+1}$, $\mathfrak{p}_j \in \mathcal{O}^n$ and $\mathfrak{q} \in \mathcal{O}^m$. Observing that $Y_\lambda \in \mathcal{R}(m;\mathbb{F})$ and Proposition I.4.8 we have

$$\det Y_\lambda[Q] \geq \alpha \prod_{j=1}^{m} Y_\lambda[q_j] = \alpha \prod_{j=1}^{m} (Y[p_j] + \lambda N(q_j)) ,$$

where $\alpha = c(m;\mathbb{F})^{-1}$. From $p_j \neq 0$ we obtain $Y[p_j] \geq \mu(Y) = y_1 \geq \frac{1}{2}\sqrt{3}$. Suppose that $q \neq 0$. Since Q is part of a unimodular matrix, it follows that

$$\det Y_\lambda[Q] \geq \alpha(\tfrac{1}{2}\sqrt{3})^{m-1} \lambda .$$

Clearly it suffices to choose $\lambda_o = c(m;\mathbb{F})\left(\frac{2}{\sqrt{3}}\right)^{m-1}$ in this case.

Otherwise suppose that $q = 0$. We get $\operatorname{rank} Q = m < n+1$ and we can find a matrix $V_o = (Q,*) \in GL(n;\mathcal{O})$, if we pay attention to Theorem I.2.3. Put

$$C_o = \begin{pmatrix} C_1 & 0 \\ 0 & 0 \end{pmatrix} \bar{V}_o' , \quad D_o = \begin{pmatrix} D_1 & 0 \\ 0 & I \end{pmatrix} V_o^{-1} , \quad C_o, D_o \in \operatorname{Mat}(n;\mathcal{O}) .$$

Then $(C_o\ D_o)$ turns out to be the second row of a matrix $M_o \in \Gamma(n;\mathbb{F})$ in view of Theorem 2.5. Because of $Z \in F(n;\mathbb{F})$ one computes that

$$|\det \widehat{M\{Z_\lambda\}}| = |\det \widehat{M_o\{Z\}}| \geq 1 ,$$

hence $Z_\lambda \in F(n+1;\mathbb{F})$.

□

<u>Corollary 3.12.</u> <u>The quantity</u>

$$s(n;\mathbb{F}) := \sup\,\{\operatorname{tr}(Y^{-1}) \; ; \; Y \in F(n;\mathbb{F})\}$$

<u>is finite</u>. <u>One has</u> $s(1;\mathbb{F}) = \frac{2}{3}\sqrt{3}$ <u>and</u>

$$s(n;\mathbb{F}) \leq s(n+1;\mathbb{F}) .$$

<u>Proof.</u> $s(n;\mathbb{F}) \leq n\rho^{-1} < \infty$ holds in view of Proposition 3.1. One computes $s(1;\mathbb{F}) = \frac{2}{3}\sqrt{3}$ from the explicit description of $F(1;\mathbb{F})$. Given $Z \in F(n;\mathbb{F})$ then $Z_\lambda = [Z, i\lambda] \in F(n+1;\mathbb{F})$ holds for sufficiently large λ . Now

$$\operatorname{tr}(Y_\lambda^{-1}) = \frac{1}{\lambda} + \operatorname{tr}(Y^{-1})$$

completes the proof.

□

What about the cases $n = 1$ resp. $n = 2$? It is well-known that

$$F(1;\mathbb{F}) = \{z = x + iy \in \mathbb{C} \; ; \; -\tfrac{1}{2} \leq x \leq \tfrac{1}{2} , \; x^2 + y^2 \geq 1\} .$$

In the case $n = 2$ one can easily derive estimations for the quantities $s(2;\mathbb{F})$. Especially one has

$$s(2;\mathbb{R}) \leq \frac{16}{9}\sqrt{3} , \quad s(2;\mathbb{F}) \leq \frac{8}{3}\sqrt{3} \quad \text{for } \mathbb{F} = \mathbb{C}, \mathbb{H} .$$

§4 Congruence subgroups

Beside the modular group congruence subgroups play an essential role in the theory of modular forms. Congruence subgroups of SIEGEL's modular group were examined by KOECHER [35]. The case of the Hermitian modular group was treated by BRAUN [8].

Given a two-sided ideal $\mathcal{I}$ in $\mathcal{O}$ and $P,Q \in \mathrm{Mat}(m,n;\mathbb{F})$ we define

$$P \equiv Q \bmod \mathcal{I}\ , \text{ if } \ P - Q \in \mathrm{Mat}(m,n;\mathcal{I})\ .$$

Considering $P,Q \in \mathrm{Mat}(n;\mathcal{O})$ this equivalence relation is compatible with the ring structure of $\mathrm{Mat}(n;\mathcal{O})$.

On the other hand notice that each two-sided ideal $\mathcal{I} \neq \{0\}$ is generated by an invariant element in view of Proposition I.1.9. Again let $S(\mathbb{F}) := E(\mathbb{F}) \cap \mathrm{Cent}\,\mathbb{F}$, hence

$$S(\mathbb{R}) = S(\mathbb{H}) = \{\pm 1\}\ ,\ S(\mathbb{C}) = \{\pm e_1, \pm e_2\}\ .$$

Given $q \in \mathbb{N}$ put

$$\Gamma(n;\mathbb{F})[q] := \{M \in \Gamma(n;\mathbb{F})\ ;\ M \equiv \varepsilon I \bmod q\mathcal{O}\ ,\ \text{where } \ \varepsilon \in S(\mathbb{F})\}\ .$$

<u>Proposition 4.1.</u> <u>$\Gamma(n;\mathbb{F})[q]$ is a normal subgroup of $\Gamma(n;\mathbb{F})$ of finite index.</u>

<u>Proof.</u> Clear, since the index may roughly be estimated by $q^{(2rn)^2}$.

□

The index of SIEGEL's modular group was explicitly computed by KOECHER [35].

<u>Definition.</u> $\Gamma(n;\mathbb{F})[q]$ is called the <u>principal congruence subgroup of level</u> q . By a <u>congruence subgroup of level</u> q we mean a subgroup of $\Gamma(n;\mathbb{F})$ containing $\Gamma(n;\mathbb{F})[q]$. The congruence subgroup

$$\Gamma(n;\mathbb{F})\langle q\rangle := \{M = \begin{pmatrix} A & B \\ C & D \end{pmatrix} \in \Gamma(n;\mathbb{F})\ ;\ C \equiv 0 \bmod q\mathcal{O}\ \}$$

is called <u>theta-group of level</u> q .

Theorem 4.2. Given $q \in \mathbb{N}$ the theta-group of level q $\Gamma(n;\mathbb{F})\langle q\rangle$ is generated by the matrices

$$\begin{pmatrix} \bar{U}' & 0 \\ 0 & U^{-1} \end{pmatrix}, \begin{pmatrix} I & S \\ 0 & I \end{pmatrix}, \begin{pmatrix} I & 0 \\ qS & I \end{pmatrix}, \quad I \times K ,$$

where $U \in GL(n;\mathcal{O})$, $S \in Sym(n;\mathcal{O})$ and $K \in \Gamma(1;\mathbb{F})\langle q\rangle$.

Proof. One has $\Gamma(n;\mathbb{F})\langle 1\rangle = \Gamma(n;\mathbb{F})$. Because of

$$\begin{pmatrix} I & I \\ 0 & I \end{pmatrix} \begin{pmatrix} I & 0 \\ -I & I \end{pmatrix} \begin{pmatrix} I & I \\ 0 & I \end{pmatrix} = \begin{pmatrix} 0 & I \\ -I & 0 \end{pmatrix} = J$$

the assertion in the case $q = 1$ follows from Theorem 2.3.

Now suppose that $q > 1$. We use induction on n , where the case $n = 1$ is trivial. Given $n > 1$ let Δ_n denote the subgroup of $\Gamma(n;\mathbb{F})\langle q\rangle$ generated by the matrices quoted above.

Let $M = \begin{pmatrix} A & B \\ C & D \end{pmatrix} \in \Gamma(n;\mathbb{F})\langle q\rangle$. Replacing M by $\begin{pmatrix} \bar{U}' & 0 \\ 0 & U^{-1} \end{pmatrix} M$, $U \in GL(n;\mathcal{O})$, we may assume that the first column of A possesses the form $(a,0,\ldots,0)'$ in view of Corollary I.2.4. One has $a \neq 0$, as otherwise $q > 1$ would be a common right divisor of the first column of M in contradiction to Lemma I.2.8. Let $(c_1,\ldots,c_n)'$ denote the first column of C and suppose that u is a $\text{gcrd}(a,c_1)$. Thus there exist $\alpha,\gamma \in \mathcal{O}$ satisfying $a = \alpha u$, $c_1 = \gamma u$. Lemma 1.1 yields $\bar{A}'C = \bar{C}'A$, hence $\bar{a}c_1 = \bar{c}_1 a \in \mathbb{Z}$. Now $u \neq 0$ implies $\bar{\alpha}\gamma = \bar{u}^{-1}\,\bar{a}c_1 u^{-1} = N(u)^{-1}\,\bar{a}c_1 = \bar{\gamma}\alpha$. Because of the choice of u the elements α,γ are relatively right-prime. We obtain a matrix $K \in \Gamma(1;\mathbb{F})$ having $\begin{pmatrix} \alpha \\ \gamma \end{pmatrix}$ as its first column according to Corollary 2.4. Without restriction we may assume that $K \in Mat(2;\mathbb{Z})$ in view of Lemma 2.6. Since a and q are relatively right-prime, u and q have the same property. From $q^{-1}c_1 = q^{-1}\gamma u \in \mathcal{O}$ and $q,\gamma \in \mathbb{Z}$ it follows that $q^{-1}\gamma \in \mathbb{Z}$, hence $K \in \Gamma(1;\mathbb{F})\langle q\rangle$ as well as $K^{-1} \in \Gamma(1;\mathbb{F})\langle q\rangle$. Define $\pi \in \gamma_n$ by $\pi(1) = n$, $\pi(n) = 1$ and $\pi(j) = j$, $1 < j < n$. If $P = P_\pi$ denotes the corresponding permutation matrix, one computes

$$K^{-1} \times I = \begin{pmatrix} P & 0 \\ 0 & P \end{pmatrix} (I \times K^{-1}) \begin{pmatrix} P & 0 \\ 0 & P \end{pmatrix} \in \Delta_n .$$

Thus $(u,c_2,\ldots,c_n,0,c_2,\ldots,c_n)'$ turns out to be the first column of

$$M_1 = \begin{pmatrix} I & I \\ 0 & I \end{pmatrix} (K^{-1} \times I)\, M .$$

According to Corollary 2.4 $(u,c_2,\ldots,c_n)$ are relatively right-prime, and hence the first column of a matrix $U \in GL(n;\mathcal{O})$ in view of Lemma I.2.8. It follows that $U^{-1}(u,c_2,\ldots,c_n)' = e_1$. Next Lemma 1.1 implies that we can find an $S \in Sym(n;\mathcal{O})$ such that $(1,0,\ldots,0)'$ is the first column of

$$M_2 = \begin{pmatrix} I & 0 \\ qS & I \end{pmatrix} \begin{pmatrix} U^{-1} & 0 \\ 0 & \bar{U}' \end{pmatrix} M_1 .$$

We obtain from the fundamental relations in Lemma 1.1 that M possesses the form

$$M_2 = \begin{pmatrix} 1 & \bar{a}_1' & b & \bar{h}_1' \\ 0 & A_0 & h_2 & B_0 \\ 0 & 0 & 1 & 0 \\ 0 & C_0 & \hat{a}_2 & D_0 \end{pmatrix}, \text{ where } M_0 = \begin{pmatrix} A_0 & B_0 \\ C_0 & D_0 \end{pmatrix} \in \Gamma(n-1;\mathbb{F})\langle q\rangle .$$

The induction hypothesis yields $M_0^{-1} \in \Delta_{n-1}$ and $I^{(2)} \times M_0^{-1} \in \Delta_n$. Therefore we have

$$M_3 = (I^{(2)} \times M_0^{-1})M_2 = \begin{pmatrix} * & * \\ 0 & * \end{pmatrix}, \text{ where } 0 = 0^{(n)} .$$

It follows from Proposition 1.2 that $M_3 \in \Delta_n$ and hence $M \in \Delta_n$. □

Given a congruence subgroup C of $\Gamma(n;\mathbb{F})$ the index $[\Gamma(n;\mathbb{F}) : C] = l$ is finite. Let $M_1,\ldots,M_l$ denote a complete set of representatives of the right cosets, i.e.

(1) $$\Gamma(n;\mathbb{F}) = \bigcup_{j=1}^{l} C\, M_j .$$

In the case $n = 1$ consider only $\mathbb{F} = \mathbb{R}$ and define

(2) $$F(C) := \bigcup_{j=1}^{l} M_j \langle F(n;\mathbb{F})\rangle ,$$

where the definition depends on the special choice of the representatives.

<u>Theorem 4.3.</u> <u>Given a congruence subgroup C of $\Gamma(n;\mathbb{F})$ then $F(C)$ is a fundamental domain of the half-space $H(n;\mathbb{F})$ with respect to the action of C, where the symplectic volume of $F(C)$ is finite.</u>

<u>Proof.</u> $F(C)$ is closed and contains interior points, since $F(n;\mathbb{F})$ has the same properties and symplectic transformations are topological maps. Given $Z \in H(n;\mathbb{F})$ there is an $M \in \Gamma(n;\mathbb{F})$ satisfying $M\langle Z\rangle \in F(n;\mathbb{F})$. According to (1) we can find $K \in C$ and j, $1 \leqq j \leqq l$, such that $M^{-1} = K M_j$ or $K^{-1} = M_j M$, hence

$$K^{-1}\langle Z\rangle = M_j\langle M\langle Z\rangle\rangle \in F(C) .$$

Let $K \in C$ and Z such that Z and $K\langle Z\rangle$ belong to the interior of $F(C)$. Thus $K\langle F(C)\rangle \cap F(C)$ contains interior points. Since the union (2) is finite, we can find j,k, $1 \leqq j,k \leqq l$, such that $K M_j\langle F(n;\mathbb{F})\rangle \cap M_k\langle F(n;\mathbb{F})\rangle$ and therefore $M_k^{-1}KM_j\langle F(n;\mathbb{F})\rangle \cap F(n;\mathbb{F})$ possess interior points. As $F(n;\mathbb{F})$ is a fundamental domain, it follows that $M_k^{-1}KM_j = \varepsilon I$, where $\varepsilon \in S(\mathbb{F})$, if we consider only $\mathbb{F} = \mathbb{R}$, whenever $n = 1$. Hence one has $\bar{\varepsilon}K = M_k M_j^{-1} \in C$, thus $M_j = M_k$ and $K = \varepsilon I$.

$F(C)$ has only a finite number of neighbors, since $F(n;\mathbb{F})$ has the same property. Given a compact subset of $H(n;\mathbb{F})$ only finitely many images under modular transformations meet $F(C)$.

Finally, one has in view of Proposition 3.9 and Theorem 1.10 that

$$\operatorname{vol}(C) := \int_{F(C)} dv = l \cdot \int_{F(n;\mathbb{F})} dv$$

$$= l \cdot \operatorname{vol}(\Gamma(n;\mathbb{F})) < \infty ,$$

if we consider only $\mathbb{F} = \mathbb{R}$ in the case $n = 1$. □

But another important property gets lost if we consider congruence subgroups. In contradiction to the full modular group (cf. Proposition 3.1), in general there does not exist $\rho = \rho(C)$ satisfying

$$\operatorname{Im} Z \geqq \rho I$$

for all $Z \in F(C)$, if C denotes a proper congruence subgroup of $\Gamma(n;\mathbb{F})$.

Chapter III Modular forms

This chapter deals with the vector space of modular forms. The fundamental theorems of finiteness are proved. Especially we will show that the dimension of the vector space of modular forms of weight k grows as k^h at the very most, whenever k tends to ∞, where $h = n + \frac{1}{2}rn(n-1) = \dim \mathrm{Sym}(n;\mathbb{F})$.

In §1 we will examine analytic class invariants and derive a FOURIER-expansion for this kind of functions. Since modular forms especially are analytic class invariants, the results can be applied.

In §2 we estimate the dimension of the vector space of modular forms. By means of SIEGEL's ϕ-operator the examination can be reduced to cusp forms. As a consequence we show that there are at most $h+2$ algebraically independent modular forms.

In §3 we consider modular forms with respect to congruence subgroups. The results of the preceding paragraph can easily be applied to this larger class of modular forms.

Finally, in §4 we examine relations between modular forms on SIEGEL's, Hermitian half-space and the half-space of quaternions. Especially the FOURIER-expansion of the restricted modular form can easily be derived from that of the given modular form.

§1 Analytic class invariants

In this paragraph we shall deduce a FOURIER-expansion for analytic class invariants and we shall show that KOECHER's principle is valid. The results are applied to modular forms considered as special analytic class invariants. Finally the growth of FOURIER-coefficients of modular forms is estimated.

The results on the theory of class invariants lean on KOECHER [37], II,§2 .

The dual lattice $\mathcal{O}^{\#}$ of the ordering $\mathcal{O}$ with respect to the canonical scalar product has been determined in Proposition I.1.14, namely

$$\mathcal{O}^{\#}(\mathbb{F}) = \mathcal{O}(\mathbb{F}) \quad \text{for} \quad \mathbb{F} = \mathbb{R},\mathbb{C} ,$$

$$\mathcal{O}^{\#}(\mathbb{H}) = \mathbb{Z}2e_1 + \mathbb{Z}(e_1+e_2) + \mathbb{Z}(e_1+e_3) + \mathbb{Z}(e_1+e_4) .$$

We recall the definition of

$$\tau : \mathrm{Sym}(n;\mathbb{F}) \times \mathrm{Sym}(n;\mathbb{F}) \longrightarrow \mathbb{R} , \quad \tau(S,T) := \tfrac{1}{2}\,\mathrm{trace}\,(ST+TS) .$$

Let $\mathrm{Sym}(n;\mathcal{O})$ again denote the lattice of integral Hermitian matrices. We determine the dual lattice

$$\mathrm{Sym}^{\tau}(n;\mathcal{O}) := \{T \in \mathrm{Sym}(n;\mathbb{F}) \; ; \; \tau(S,T) \in \mathbb{Z} \quad \text{for all} \quad S \in \mathrm{Sym}(n;\mathcal{O})\}$$

of $\mathrm{Sym}(n;\mathcal{O})$ with respect to τ .

<u>Lemma 1.1.</u> <u>Given</u> $T = (t_{jk}) \in \mathrm{Sym}(n;\mathbb{F})$ <u>the following assertions are equivalent</u>:

(i) $T \in \mathrm{Sym}^{\tau}(n;\mathcal{O})$.

(ii) $T[\mathfrak{g}] \in \mathbb{Z}$ <u>for all</u> $\mathfrak{g} \in \mathcal{O}^n$.

(iii) $t_j \in \mathbb{Z}$, $1 \leq j \leq n$, $2t_{jk} \in \mathcal{O}^{\#}$, $1 \leq j < k \leq n$.

<u>Proof.</u> "(i) ⇒ (ii)" We observe that $T[\mathfrak{g}] = \tau(T,\mathfrak{g}\bar{\mathfrak{g}}')$ and $\mathfrak{g}\bar{\mathfrak{g}}' \in \mathrm{Sym}(n;\mathcal{O})$ for all $\mathfrak{g} \in \mathcal{O}^n$.

"(ii) ⇒ (iii)" Choosing $\mathfrak{g} = \mathfrak{e}_j$ leads to $t_j = T[\mathfrak{e}_j] \in \mathbb{Z}$ for $1 \leq j \leq n$. Putting $\mathfrak{g} = g\mathfrak{e}_j + \mathfrak{e}_k$, $g \in \mathcal{O}$, $j \neq k$, yields

$$T[\mathfrak{g}] = t_j N(g) + t_k + 2\,\mathrm{Re}(\bar{g}\,t_{jk}) \in \mathbb{Z} ,$$

hence $2\,\mathrm{Re}(\bar{g}\,t_{jk}) \in \mathbb{Z}$ for all $g \in \mathcal{O}$. Thus we have $2t_{jk} \in \mathcal{O}^{\#}$.

"(iii) ⇒ (i)" We notice that

$$\tau(S,T) = \sum_{j=1}^{n} s_j t_j + \sum_{1\le j<k\le n} 2\,\mathrm{Re}(\bar{s}_{jk} t_{jk}) \,. \qquad \square$$

Given a lattice L let vol L denote the Euclidean volume of a fundamental parallelotope of L. Especially one has

$$\begin{aligned} \mathrm{vol}(\mathcal{O}(\mathbb{F})) &= 1 \quad \text{for } \mathbb{F} = \mathbb{R},\mathbb{C}\,, \\ \mathrm{vol}(\mathcal{O}(\mathbb{H})) &= \tfrac{1}{2}\,, \\ \mathrm{vol}(\mathcal{O}^{\#}) &= (\mathrm{vol}\ \mathcal{O})^{-1}\,, \\ \mathrm{vol}\ \mathrm{Sym}(n;\mathcal{O}) &= (\mathrm{vol}\ \mathcal{O})^{\frac{1}{2}n(n-1)}\,. \end{aligned}$$

We call a function $\varphi : H(n;\mathbb{F}) \longrightarrow \mathbb{C}$ <u>periodic</u> if it satisfies

$$\varphi(Z+S) = \varphi(Z)$$

for all $Z \in H(n;\mathbb{F})$ and $S \in \mathrm{Sym}(n;\mathcal{O})$. Let $C(n;\mathbb{F})$ again denote the unit cube in $\mathrm{Sym}(n;\mathbb{F})$ defined in II,§3.

<u>Theorem 1.2.</u> <u>If</u> $\varphi : H(n;\mathbb{F}) \longrightarrow \mathbb{C}$ <u>is holomorphic and periodic, then</u> φ <u>possesses an absolutely convergent</u> FOURIER-<u>series</u>

$$\varphi(Z) = \sum_T \alpha(T)\, e^{2\pi i \tau(T,Z)}\,, \quad Z \in H(n;\mathbb{F})\,,$$

<u>where</u> T <u>runs through</u> $\mathrm{Sym}^{\tau}(n;\mathcal{O})$. <u>The</u> FOURIER-<u>coefficients</u> $\alpha(T)$ <u>do not depend on</u> Z, <u>are uniquely determined by</u> φ <u>and given by</u>

$$\alpha(T) = (\mathrm{vol}\ \mathrm{Sym}(n;\mathcal{O}))^{-1}\, e^{2\pi\tau(T,Y)} \int_{C(n;\mathbb{F})} \varphi(X+iY)\, e^{-2\pi i\,\tau(T,X)}\, dX\,,$$

<u>where</u> $Y > 0$ <u>is fixed and</u> dX <u>denotes the Euclidean volume element of</u> $\mathrm{Sym}(n;\mathbb{F})$.

<u>Proof.</u> Let the map $\rho: \mathrm{Sym}(n;\mathbb{F}) \longrightarrow \mathbb{R}^h$, $h = \dim \mathrm{Sym}(n;\mathbb{F})$, denote the projection onto the components with respect to the canonical basis, i.e. given $X = (x_{kl}) \in \mathrm{Sym}(n;\mathbb{F})$, where $x_{kl} = \sum_{j=1}^{r} x_{kl}^{(j)} e_j$ for $k \neq l$, then put $\rho(X) := (x_1,\ldots,x_n,x_{12}^{(1)},\ldots,x_{12}^{(r)},\ldots,x_{n-1n}^{(r)})' \in \mathbb{R}^h$. Define $P := P(\mathbb{F}) \in \mathrm{Mat}(r;\mathbb{R})$ by $P(\mathbb{F}) := I^{(r)}$ for $\mathbb{F} = \mathbb{R},\mathbb{C}$ and

$$P(\mathbb{H}) := \begin{pmatrix} 1 & 0 & 0 & -1 \\ 0 & 1 & 0 & -1 \\ 0 & 0 & 1 & -1 \\ 0 & 0 & 0 & 2 \end{pmatrix}.$$

Additionally put

$$Q := [I,P,\dots,P] \in \mathrm{Mat}(h;\mathbb{R}) \quad , \; I = I^{(n)} \; .$$

Thus an inspection yields $Q\rho(C(n;\mathbb{F})) = [-\frac{1}{2};\frac{1}{2}]^h$ and $Q\rho(\mathrm{Sym}(n;\mathcal{O})) = \mathbb{Z}^h$. The function φ is related to a function ψ defined on the open subset $\mathcal{U} := Q\rho(\mathrm{Sym}(n;\mathbb{F})) + iQ\rho(\mathrm{Pos}(n;\mathbb{F}))$ of $\mathbb{C}^h$ by

$$\psi(\mathfrak{u}+i\mathfrak{v}) := \varphi\left(\rho^{-1}(Q^{-1}\mathfrak{u}) + i\rho^{-1}(Q^{-1}\mathfrak{v})\right) .$$

In view of the construction ψ turns out to be holomorphic and periodic with period 1 in each of the h complex variables. According to [58], Cor. 1.8 and 1.9, p.249, ψ possesses an absolutely convergent FOURIER-expansion:

$$\psi(\mathfrak{u}+i\mathfrak{v}) = \sum_{\mathfrak{g}\in\mathbb{Z}^m} \beta(\mathfrak{g})\, e^{2\pi i\mathfrak{g}'\mathfrak{u}} \quad , \; \mathfrak{u}+i\mathfrak{v} \in \mathcal{U} \; .$$

One computes $\mathfrak{g}'\mathfrak{u} = \tau(\rho^{-1}(R\mathfrak{g}),\rho^{-1}(Q\mathfrak{u}))$, where $R = [I,\frac{1}{2}P'^{-1},\dots,\frac{1}{2}P'^{-1}]$ and $\rho^{-1}(R\mathbb{Z}^h) = \mathrm{Sym}^{\tau}(n;\mathcal{O})$. The FOURIER-coefficient $\beta(\mathfrak{g})$ is given by

$$\beta(\mathfrak{g}) = \int_{[-\frac{1}{2};\frac{1}{2}]^h} \psi(\mathfrak{u}+i\mathfrak{v})\, e^{-2\pi i\mathfrak{g}'\mathfrak{u}}\, d\mathfrak{u} \; .$$

Hence we obtain the quoted FOURIER-expansion of f, where

$$\alpha(T;Y) = e^{2\pi\tau(T,Y)}\,(\det Q)^{-1} \int_{C(n;\mathbb{F})} \varphi(X+iY)\, e^{-2\pi i\tau(T,X)}\, dX \; .$$

We have $\det Q = \mathrm{vol}\,\mathrm{Sym}(n;\mathcal{O})$. If z denotes a component of Z, then $\frac{\partial\varphi}{\partial\bar{z}}$ turns out to be periodic again, and we obtain

$$\frac{\partial\varphi}{\partial\bar{z}} = \sum_T \frac{\partial\alpha(T;Y)}{\partial\bar{z}}\, e^{2\pi i\tau(T,Z)} \quad , \; Z \in H(n;\mathbb{F}) \; .$$

But one has $\frac{\partial\varphi}{\partial\bar{z}} \equiv 0$, because φ is holomorphic. The uniqueness of the FOURIER-expansion yields $\frac{\partial\alpha(T;Y)}{\partial\bar{z}} = 0$. Thus the FOURIER-coefficients do not depend on Y.

□

Given $T \geq 0$ then $\det T \geq 0$ holds. Hence $(\det T)^{\alpha}$ is well-defined for any $\alpha \geq 0$.

<u>Proposition 1.3.</u> <u>Given</u> $\alpha \geq 0$ <u>and</u> $\beta > 0$ <u>the series</u>

$$\sum_{T \in \mathrm{Sym}^{\tau}(n;\mathcal{O}),\, T \geq 0} (\det T)^{\alpha}\, e^{-\beta \operatorname{tr} T}$$

<u>converges</u>.

<u>Proof.</u> Given $T \in \mathrm{Sym}^{\tau}(n;\mathcal{O})$, $T \geq 0$, then $\operatorname{tr} T \in \mathbb{N}_0$ holds. There are at most $(4t+1)^h$ matrices $T \in \mathrm{Sym}^{\tau}(n;\mathcal{O})$ satisfying $T \geq 0$ and $\operatorname{tr}(T) = t$. These matrices additionally fulfill $\det T \leq t^n$ in view of HADAMARD's inequality. The convergence of the series

$$\sum_{t=0}^{\infty} (4t+1)^h\, t^{\alpha n}\, e^{-\beta t}$$

completes the proof.

□

Let $SL(n;\mathcal{O}) := \{U \in GL(n;\mathcal{O})\ ;\ \det U = 1\}$ for $\mathbb{F} = \mathbb{R}, \mathbb{C}$ and $SL(n;\mathcal{O}(\mathbb{H})) := GL(n;\mathcal{O}(\mathbb{H}))$.

<u>Definition.</u> A function $\varphi : H(n;\mathbb{F}) \longrightarrow \mathbb{C}$ is called an <u>analytic class invariant</u> if the following conditions are satisfied:

(I.1) φ is holomorphic.

(I.2) φ is periodic.

(I.3) $\varphi(Z[U]) = \varphi(Z)$ for all $Z \in H(n;\mathbb{F})$ and $U \in SL(n;\mathcal{O})$.

Obviously, the set of analytic class invariants forms a $\mathbb{C}$-algebra. Next we construct an example.

<u>Lemma 1.4.</u> <u>Given</u> $T \in \mathrm{Sym}^{\tau}(n;\mathcal{O})$, $T \geq 0$, <u>then</u>

$$\omega_T(Z) := \sum_{S \sim T} e^{2\pi i \tau(S,Z)}, \quad Z \in H(n;\mathbb{F}),$$

<u>becomes an analytic class invariant, where</u> $S \sim T$ <u>means that</u> S <u>runs through the set</u> $\{T[U]\ ;\ U \in SL(n;\mathcal{O})\}$.

<u>Proof.</u> Given $Z = X + iY \in H(n;\mathbb{F})$ there exists $\beta > 0$ satisfying $Y \geq \beta I$. Now $T \geq 0$ implies $S \geq 0$ and $\tau(S,Y) \geq \beta \operatorname{tr}(S)$ for all $S \sim T$. Thus we compute that

$$\left| e^{2\pi i \tau(S,Z)} \right| = e^{-2\pi\tau(S,Y)} \leq e^{-2\pi\beta \operatorname{tr}(S)} .$$

The series

$$\sum_{S\sim T} e^{-2\pi\beta\,\mathrm{tr}(S)}$$

converges according to Proposition 1.3, since $T \in \mathrm{Sym}^{\tau}(n;\mathcal{O})$ implies $T[U] \in \mathrm{Sym}^{\tau}(n;\mathcal{O})$ for $U \in SL(n;\mathcal{O})$. These estimations yield the locally uniform convergence of the series. Hence ω_T is holomorphic. We obtain $\tau(S,R) \in \mathbb{Z}$ for all $R \in \mathrm{Sym}(n;\mathcal{O})$ from $S \in \mathrm{Sym}^{\tau}(n;\mathcal{O})$. Thus ω_T is periodic. The property $\omega_T(Z[U]) = \omega_T(Z)$ for $U \in SL(n;\mathcal{O})$ follows from $\tau(S,Z[U]) = \tau(S[\bar{U}'],Z)$ by rearranging the series.

□

Thus we are able to prove the important

<u>Lemma 1.5.</u> KOECHER'<u>s principle</u>

<u>Suppose that the series</u> $\sum_{S\sim T} e^{2\pi i\tau(S,Z)}$, $Z \in H(n;\mathbb{F})$, <u>converges absolutely for some</u> $T \in \mathrm{Sym}^{\tau}(n;\mathcal{O})$. <u>If</u> $n > 1$ <u>it follows that</u> $T \geq 0$.

<u>Proof.</u> Let $T \in \mathrm{Sym}^{\tau}(n;\mathcal{O})$, which is not positive semi-definite. An argument of continuity leads to the existence of an $\mathfrak{a} \in \mathcal{O}^n$ such that $T[\mathfrak{a}] < 0$. Because of $n > 1$ we can find $\mathfrak{b} \in \mathcal{O}^n - \{0\}$ satisfying $\bar{\mathfrak{a}}'\mathfrak{b} = 0$. Given $m \in \mathbb{Z}$ then $U_m := I + m\mathfrak{a}\bar{\mathfrak{b}}' \in SL(m;\mathcal{O})$ holds in view of $U_m U_{-m} = I$ and in the cases $\mathbb{F} = \mathbb{R},\mathbb{C}$ additionally by virtue of

$$\det U_m = 1 + m\bar{\mathfrak{b}}'\mathfrak{a} = 1 .$$

Put $S_m := T[U_m]$ then $\mathrm{tr}(S_m) = \mathrm{tr}(T) + m\,\mathrm{tr}(\mathfrak{b}\bar{\mathfrak{a}}'T + T\mathfrak{a}\bar{\mathfrak{b}}') + m^2 T[\mathfrak{a}]\bar{\mathfrak{b}}'\mathfrak{b}$ holds. Because of $T[\mathfrak{a}]\bar{\mathfrak{b}}'\mathfrak{b} < 0$ the quantities $\mathrm{tr}(S_m)$ are mutually distinct and negative for some m_0 and $m \geq m_0$. Now

$$\sum_{m\geq m_0} e^{-2\pi\,\mathrm{tr}(S_m)}$$

is a partial series of $\omega_T(iI)$, which diverges because of $e^{-2\pi\,\mathrm{tr}(S_m)} \geq 1$

□

As a consequence we derive

<u>Theorem 1.6.</u> <u>Let</u> $n > 1$ <u>and</u> φ <u>be an analytic class invariant</u>. <u>In the</u> FOURIER-<u>expansion</u>

$$\varphi(Z) = \sum_{T} \alpha(T)\, e^{2\pi i \tau(T,Z)} \quad , \quad Z \in H(n;\mathbb{F}) \; ,$$

all FOURIER-coefficients $\alpha(T)$ vanish, whenever T is not positive semi-definite. There exists a representation

$$\varphi(Z) = \sum_{\{T\} \geq 0} \alpha(T)\, \omega_T(Z) \quad , \quad Z \in H(n;\mathbb{F}) \; ,$$

where the sum is taken over a set of representatives of the orbits $\{T[U] \; ; \; U \in SL(n;\mathcal{O})\}$. Given $\alpha > 0$ then φ is bounded in the domain $\{Z \in H(n;\mathbb{F}) \; ; \; \text{Im}\, Z \geq \alpha I\}$.

Proof. First $\alpha(T) = \alpha(T[U])$ for $U \in SL(n;\mathcal{O})$ follows from (I.3) and the uniqueness of the FOURIER-expansion of φ . Because of the absolute convergence it is allowed to rearrange the FOURIER-series and to collect the terms of the orbits $\{T[U] \; ; \; U \in SL(n;\mathcal{O})\}$. If $\alpha(T) \neq 0$, one has $T \geq 0$ in view of the preceding lemma.

Since the FOURIER-series converges in $Z = i\tfrac{1}{2}\beta I$, there exists $C > 0$ satisfying

$$\left|\alpha(T)\, e^{2\pi i \tau(T,Z)}\right| = |\alpha(T)|\, e^{-\pi\beta\, \text{tr}(T)} \leq C$$

for all $T \in \text{Sym}^{\tau}(n;\mathcal{O})$, $T \geq 0$. If $Y \geq \beta I$ one has $\tau(T,Y) \geq \beta\, \text{tr}(T)$. Hence we estimate

$$\begin{aligned} |\varphi(Z)| &= \left| \sum_{T \geq 0} \alpha(T)\, e^{2\pi i \tau(T,Z)} \right| \\ &\leq \sum_{T \geq 0} |\alpha(T)|\, e^{-2\pi\tau(T,Y)} \\ &\leq C \sum_{T \geq 0} e^{-\pi\beta\, \text{tr}(T)} \end{aligned}$$

for $Z = X + iY \in H(n;\mathbb{F})$ and $Y \geq \beta I$. The last series converges according to Proposition 1.3.

□

Now we are going to explain the notion of a modular form. The set $\text{Sym}(n;\mathbb{C}) \otimes_{\mathbb{R}} \mathbb{C}$ can be identified with $\text{Mat}(n;\mathbb{C})$ in a canonical way and $GL(n;\mathbb{C})$ coincides with the set of invertible elements in $\text{Sym}(n;\mathbb{C}) \otimes_{\mathbb{R}} \mathbb{C}$. Given $M \in Sp(n;\mathbb{C})$ and $Z \in H(n;\mathbb{C})$ then $M\{Z\} = CZ + D$ belongs to $GL(n;\mathbb{C})$ according to Theorem II.1.7. Hence $(\det M\{Z\})^k$ is well-defined for any $k \in \mathbb{Z}$. Given $M \in Sp(n;\mathbb{H})$, $Z \in H(n;\mathbb{H})$ and an even $k \in \mathbb{Z}$ we put

$$(\det M\{Z\})^k := (\det \check{M}\{\check{Z}\})^{\frac{1}{2}k} .$$

Because of $\check{M} \in Sp(2n;\mathbb{C})$ and $\check{Z} \in H(2n;\mathbb{C})$ this definition makes sense.

Given a function $f: H(n;\mathbb{F}) \longrightarrow \mathbb{C}$, $M \in Sp(n;\mathbb{F})$ and $k \in \mathbb{Z}$, which is supposed to be even in the case $\mathbb{F} = \mathbb{H}$, we define

$$f|_k M : H(n;\mathbb{F}) \longrightarrow \mathbb{C} , \; Z \longmapsto (\det M\{Z\})^{-k} f(M\langle Z\rangle) .$$

Let $M,N \in Sp(n;\mathbb{F})$. It follows from Theorem II.1.7 that

(1) $$f|_k M|_k N = f|_k MN .$$

<u>Definition.</u> Let $k \in \mathbb{Z}$, where $k \equiv 0 \bmod 2$ in the case $\mathbb{F} = \mathbb{H}$. A function $f: H(n;\mathbb{F}) \longrightarrow \mathbb{C}$ is called a <u>modular form of weight</u> k <u>on</u> $H(n;\mathbb{F})$ if the following conditions are fulfilled:

(M.1) f is holomorphic.

(M.2) $f|_k M = f$ for all $M \in \Gamma(n;\mathbb{F})$.

(M.3) In the case $n = 1$ the function f is bounded in the domain $\{z \in \mathbb{C} ; \operatorname{Im} z \geq \beta\}$ for any $\beta \in \mathbb{R}$, $\beta > 0$.

We call f a <u>modular form of degree</u> n and define

$$[\Gamma(n;\mathbb{F}),k] := \{f \; ; \; f \text{ is modular form of weight } k \text{ on } H(n;\mathbb{F})\}.$$

Clearly, $[\Gamma(n;\mathbb{F}),k]$ is a $\mathbb{C}$-vector space. All constant functions are modular forms of weight 0. Given $f \in [\Gamma(n;\mathbb{F}),k]$ and $g \in [\Gamma(n;\mathbb{F}),l]$ one has

$$fg \in [\Gamma(n;\mathbb{F}),k+l] .$$

Considering Lemma II.2.6 we notice that

$$[\Gamma(1;\mathbb{H}),k] = [\Gamma(1;\mathbb{R}),k] \quad , \text{ if } k \equiv 0 \bmod 2 ,$$

$$[\Gamma(1;\mathbb{C}),k] = \begin{cases} [\Gamma(1;\mathbb{R}),k] & , \text{ if } k \equiv 0 \bmod 4 , \\ \{0\} & , \text{ else } . \end{cases}$$

In the case $n = 1$ we deal with usual elliptic modular forms, which are known since the last century. Nevertheless this case will always be included, since it is essential for some assertions, which are proved by induction on n.

In literature the notions of SIEGEL'<u>s</u> <u>modular forms</u> resp. <u>Hermitian modular forms</u> (with respect to the Gaussian number field) are used instead of modular forms on $H(n;\mathbb{R})$ resp. $H(n;\mathbb{C})$. The elements of $[\Gamma(n;\mathbb{H}),k]$ will be called <u>modular forms of quaternions</u>.

The modular transformations of M and εM, $\varepsilon \in S(\mathbb{F})$, (cf. I,§1) coincide. Hence

$$nk \equiv 0 \bmod s(\mathbb{F})$$

is a necessary condition for the existence of non-trivial modular forms, where $s(\mathbb{R}) = s(\mathbb{H}) = 2$ and $s(\mathbb{C}) = 4$. In this volume we shall construct non-trivial modular forms only in the case $k \equiv 0 \bmod s(\mathbb{F})$.

Lemma 1.7. *Let* $k \equiv 0 \bmod s(\mathbb{F})$. *A holomorphic function* $f : H(n;\mathbb{F}) \longrightarrow \mathbb{C}$ *satisfying condition* (M.3) *is a modular form of weight* k *if and only if the following properties are satisfied for all* $Z \in H(n;\mathbb{F})$:

(i) $f(Z+S) = f(Z)$ *for all* $S \in \mathrm{Sym}(n;\mathcal{O})$,

(ii) $f(Z[U]) = f(Z)$ *for all* $U \in GL(n;\mathcal{O})$,

(iii) $f(-Z^{-1}) = (\det Z)^k f(Z)$.

Proof. Condition (M.2) and $k \equiv 0 \bmod s(\mathbb{F})$ imply (i),(ii) and (iii). Conversely, if (i),(ii) and (iii) hold, one has $f|_k M = f$ for the generators of $\Gamma(n;\mathbb{F})$ quoted in Theorem II.2.3. Hence (M.2) follows from (1).

□

Thus modular forms are special analytic class invariants.

Corollary 1.8. *Suppose that* $k \in \mathbb{Z}$, $k \equiv 0 \bmod 2$ *in the case* $\mathbb{F} = \mathbb{H}$ *and* $f \in [\Gamma(n;\mathbb{F}),k]$.

a) f *is an analytic class invariant*.

b) f *is bounded in the fundamental domain* $\mathcal{F}(n;\mathbb{F})$ *and more generally in every domain* $\{Z \in H(n;\mathbb{F})\ ;\ \mathrm{Im}\, Z \geq \alpha I\}$, *where* $0 < \alpha \in \mathbb{R}$.

c) f *possesses a* FOURIER-*expansion*

$$f(Z) = \sum_{T \in \mathrm{Sym}^{\tau}(n;\mathcal{O}),\, T \geq 0} \alpha(T)\, e^{2\pi i \tau(T,Z)} , \quad Z \in H(n;\mathbb{F}) ,$$

where the FOURIER-*series converges absolutely and uniformly in every domain* $\{Z \in H(n;\mathbb{F})\ ;\ \mathrm{Im}\, Z \geq \alpha I\}$, *where* $0 < \alpha \in \mathbb{R}$.

Proof. It follows from (M.2) that f is an analytic class invariant. We obtain parts b) and c) in view of Theorem 1.6 and (M.3).

□

Because of the boundedness of a modular form in the fundamental domain one can estimate the growth of FOURIER-coefficients. According to KOECHER [34], Hilfssatz 13, one proves

Lemma 1.9. *Suppose that* $f \in [\Gamma(n;\mathbb{F}),k]$, $k \geq 0$, $k \equiv 0 \bmod 2$ *in the case* $\mathbb{F}=\mathbb{H}$. *Let* $\alpha(T)$ *denote the FOURIER-coefficients of* f . *Then there exists a* $C = C(f)$ *such that*

$$|\alpha(T)| \leq C\ (\det T)^{k}$$

holds for all $T \in \mathrm{Sym}^{\tau}(n;\mathcal{O})$, $T > 0$.

Proof. Given $Z = X + iY \in H(n;\mathbb{F})$ there exists an $M = \begin{pmatrix} A & B \\ C & D \end{pmatrix} \in \Gamma(n;\mathbb{F})$ satisfying $Z_1 := M\langle Z\rangle \in F(n;\mathbb{F})$. If $S \in \mathrm{Sym}(n;\mathcal{O})$, then $M_0 = \begin{pmatrix} S & -I \\ I & 0 \end{pmatrix}$ belongs to $\Gamma(n;\mathbb{F})$. We put

$$Z_0 = X_0 + iY_0 := M_0^{-1}\langle Z\rangle = (-Z+S)^{-1}$$

and obtain $Z_1 = MM_0\langle Z_0\rangle$, where $MM_0 = \begin{pmatrix} * & * \\ CS+D & -C \end{pmatrix}$.

The term $\det(\widehat{CS+D})$ becomes a non-identically vanishing polynomial in the components of S , where the degree in $s_{\nu\mu}^{(j)}$ does not exceed $2r$. Therefore we can determine $S \in \mathrm{Sym}(n;\mathcal{O})$ such that

$$(2) \qquad |x_{\nu\mu}^{(j)} - s_{\nu\mu}^{(j)}| \leq 2r \quad \text{and} \quad \det(\widehat{CS+D}) \neq 0 .$$

Since f is a modular form of weight k , we compute that

$$f(Z_0) = (\det(Z-S))^{k}\ f(Z) ,$$

$$f(Z_1) = (\det MM_0\{Z_0\})^{k}\ f(Z_0) ,$$

$$|f(Z)|^{r} = |\det(Z-S)|^{-rk}\ |\det(\widehat{CS+D})|^{-k}\ |\det(Z_0-(CS+D)^{-1}C)|^{-rk} \cdot |f(Z_1)|^{r} .$$

According to the preceding corollary f is bounded in $F(n;\mathbb{F})$. Hence there exists $C_1 = C_1(f)$ satisfying $|f(Z_1)| \leq C_1$. From $\det(\widehat{CS+D}) \in \mathbb{Z}$, $k \geq 0$ and (2) it follows that $|\det(\widehat{CS+D})|^{-k} \leq 1$. Because of $(CS+D)^{-1}C \in \mathrm{Sym}(n;\mathbb{F})$ and $|\det W| \geq \det(\mathrm{Im}\, W)$ for $W \in H(n;\mathbb{F})$ one has

$$|f(Z)| \leq C_1\ |\det(Z-S)|^{-k} \cdot (\det Y_0)^{-k} .$$

We obtain $\det Y_0 = \det Y\ |\det(Z-S)|^{-2}$ from $Z_0 = M_0^{-1}\langle Z\rangle$ and Theorem II.1.7, hence

$$|f(Z)| \leq C_1 \; |\det(Z-S)|^k \, (\det Y)^{-k} .$$

According to the choice of S in (2) the term $\det(X-S+iY)$ becomes a polynomial in $y_{\nu\mu}^{(j)}$ with bounded coefficients, when we regard Theorem I.3.4. Because of $|y_{\nu\mu}^{(j)}| \leq \sqrt{y_\nu y_\mu} \leq \operatorname{tr} Y$ there exists a suitable constant C_2 depending only on n such that

$$|\det(Z-S)| \leq C_2(1+\operatorname{tr} Y)^n ,$$

hence

$$|f(Z)| \leq C_1 C_2^k \, (1+\operatorname{tr} Y)^{nk} \, (\det Y)^{-k} .$$

Suppose that $T \in \operatorname{Sym}^\tau(n;\mathcal{O})$, $T > 0$. Because of $|\alpha(T[U])| = |\alpha(T)|$ for all $U \in GL(n;\mathcal{O})$ we may assume that $T \in \mathcal{R}(n;\mathbb{F})$. In view of Theorem I.4.12 there exists $\beta > 0$ depending only on n such that $T \geq \beta \operatorname{diag} T$ and $\operatorname{tr}(T^{-1}) \leq \beta^{-1} \operatorname{tr}\left((\operatorname{diag} T)^{-1}\right) \leq n\beta^{-1}$. Thus $|f(X+iT^{-1})| \leq C_3 \, (\det T)^k$ holds for all $X \in \operatorname{Sym}(n;\mathbb{F})$, where C_3 depends only on f. Finally Theorem 1.2 leads to

$$\begin{aligned} |\alpha(T)| &= (\operatorname{vol} \operatorname{Sym}(n;\mathcal{O}))^{-1} \, e^{2\pi\tau(T,T^{-1})} \\ &\qquad \left| \int_{C(n;\mathbb{F})} f(X+iT^{-1}) \, e^{-2\pi i\tau(T,X)} \, dX \right| \\ &\leq C_3 \, e^{2\pi n} (\det T)^k . \end{aligned}$$

□

By means of SIEGEL's ϕ-operator, which will be introduced in §2, it is also possible to estimate the growth of the FOURIER-coefficients of all positive semi-definite $T \in \operatorname{Sym}^\tau(n;\mathcal{O})$.

Next we derive a special FOURIER-expansion, whenever Z decomposes into blocks.

<u>Proposition 1.10.</u> <u>Let</u> $f \in [\Gamma(n;\mathbb{F}),k]$, $k \equiv 0 \bmod 2$ <u>in the case</u> $\mathbb{F} = \mathbb{H}$, $n = n_1 + n_2$, <u>where</u> $0 < n_1 < n$. <u>Given</u> $Z_2 \in H(n_2;\mathbb{F})$ <u>the map</u> $Z_1 \longmapsto f([Z_1,Z_2])$ <u>is a modular form of weight</u> k <u>on</u> $H(n_1;\mathbb{F})$. <u>There exists a</u> FOURIER-<u>expansion</u>

$$f([Z_1,Z_2]) = \sum_{T_2} \alpha_{T_2}(Z_1) \, e^{2\pi i\tau(T_2,Z_2)} , \quad Z_2 \in H(n_2;\mathbb{F}) ,$$

<u>where the sum is taken over</u> $T_2 \in \mathrm{Sym}^\tau(n_2;\mathcal{O})$, $T_2 \geq 0$. <u>One has</u>

$$\alpha_{T_2}(Z_1) = \sum_T \alpha(T)\, e^{2\pi i \tau(T_1,Z_1)} , \quad Z_1 \in H(n_1;\mathbb{F}) ,$$

<u>where</u> T <u>runs through the set</u> $T = \begin{pmatrix} T_1 & * \\ * & T_2 \end{pmatrix} \in \mathrm{Sym}^\tau(n;\mathcal{O})$, $T \geq 0$.

<u>Given a positive semi-definite</u> $T_2 \in \mathrm{Sym}^\tau(n_2;\mathcal{O})$ <u>then</u> $\alpha_{T_2} \in [\Gamma(n_1;\mathbb{F}),k]$.

<u>Proof.</u> Given $T = \begin{pmatrix} T_1 & * \\ * & T_2 \end{pmatrix}$ and $Z = \begin{pmatrix} Z_1 & 0 \\ 0 & Z_2 \end{pmatrix}$ then $\tau(T,Z) = \tau(T_1,Z_1) + \tau(T_2,Z_2)$. The assertion follows by a rearrangement of the FOURIER-series because of the uniqueness of the FOURIER-coefficients.

□

Clearly, the roles of Z_1 and Z_2 may be exchanged in the preceding assertion, i.e. given $Z_1 \in H(n_1;\mathbb{F})$ the map $Z_2 \longmapsto f([Z_1,Z_2])$ turns out to be a modular form of weight k on $H(n_2;\mathbb{F})$, too. Consider

$$P = \begin{pmatrix} 0 & I^{(n_2)} \\ I^{(n_1)} & 0 \end{pmatrix} , \text{ hence } \det P = (-1)^{n_1 n_2} .$$

Since $n_1 k \equiv 0 \bmod 2$ and $n_2 k \equiv 0 \bmod 2$ are necessary conditions that the arising modular forms do not vanish identically, we obtain a symmetry in the arguments

$$f([Z_1,Z_2]) = f([Z_2,Z_1]) .$$

Finally, if we consider theta-series the situation can easily be described in greater detail (cf. Lemma IV.1.14).

§2 The vector space of modular forms

In this paragraph we show that the vector space of modular forms has finite dimension. This assertion was proved by SIEGEL [54] for SIEGEL's modular forms and by BRAUN [8] for Hermitian modular forms. The proofs given below are adapted from MAASS [43], [45], §13. SIEGEL's ϕ-operator turns out to be the central tool. As a consequence we compute an upper bound for the number of algebraically independent modular forms.

Without stating explicitly the weight of modular forms of quaternions is always supposed to be even.

Sometimes we consider a formal extension to the case $n = 0$, where we define

$$[\Gamma(0;\mathbb{F}),k] := \mathbb{C} .$$

Lemma 2.1. *Given* $n \in \mathbb{N}$ *the map*

$$\phi : [\Gamma(n;\mathbb{F}),k] \longrightarrow [\Gamma(n-1;\mathbb{F}),k] , \ f \longrightarrow f|\phi ,$$

$$f|\phi\ (Z_1) := \lim_{y\to\infty} f([Z_1,iy]) , \ Z_1 \in H(n-1;\mathbb{F}) ,$$

is well-defined and linear.

Proof. Because of Corollary 1.8 the limit can be distributed through the infinite sum of the FOURIER-series. Choosing $n_1 = n-1$ in Proposition 1.10 one has $f|\phi = \alpha_0 \in [\Gamma(n-1;\mathbb{F}),k]$. Since ϕ is clearly linear, the proof is complete.

□

The homomorphism ϕ is called *SIEGEL's ϕ-operator*. The elements of $[\Gamma(n;\mathbb{F}),k]_0 := \{f \in [\Gamma(n;\mathbb{F}),k] ; f|\phi \equiv 0\} = \text{kernel } \phi$ are said to be *cusp forms*.

Proposition 2.2. *Let* $T \in \mathrm{Sym}(n;\mathbb{F})$ *be rational, i.e.* $\hat{T} \in \mathrm{Sym}(rn;\mathbb{Q})$. *If* $T \geq 0$ *and* $0 < \mathrm{rank}\, T = m < n$, *there exist* $U \in GL(n;\mathcal{O})$ *and* $T_1 \in \mathrm{Sym}(m;\mathbb{F})$ *satisfying* $T_1 > 0$ *and*

$$T[U] = \begin{pmatrix} T_1 & 0 \\ 0 & 0 \end{pmatrix} .$$

Proof. Because of rank $T < n$ there exists an $a \in \mathcal{O}^n - \{0\}$ such that $T[a] = 0$. We may suppose that a is relatively right-prime. In view of Lemma I.2.8 we can find $U \in GL(n;\mathcal{O})$ having a as its last column. Hence we compute

$$T[U] = \begin{pmatrix} S_1 & * \\ * & 0 \end{pmatrix} .$$

We obtain $T[U] = [S_1, 0]$ from $T[U] \geq 0$. An induction completes the proof.

□

The preceding assertion enables us to characterize cusp forms on the basis of their FOURIER-expansions.

Lemma 2.3. Let $f \in [\Gamma(n;\mathbb{F}),k]$. f is a cusp form if and only if f possesses a FOURIER-expansion of the form

$$f(Z) = \sum_{T \in Sym^{\tau}(n;\mathcal{O}),\, T>0} \alpha(T)\, e^{2\pi i \tau(T,Z)} \quad , \; Z \in H(n;\mathbb{F}) \; ,$$

i.e. all FOURIER-coefficients $\alpha(T)$ vanish, for which $T \in Sym^{\tau}(n;\mathcal{O})$ is not positive definite.

Proof. If f possesses the desired FOURIER-expansion, one has $f|\phi \equiv 0$, since the limit may be distributed through the infinite series.

Conversely, let f be a cusp form. In view of the proof of Lemma 2.1 one has $\alpha([T_1,0]) = 0$ for all $T_1 \in Sym^{\tau}(n-1;\mathcal{O})$ in the FOURIER-expansion. Because of $|\alpha(T[U])| = |\alpha(T)|$ for all $U \in GL(n;\mathcal{O})$ Proposition 2.2 completes the proof.

□

Lemma 2.4. Let $f \in [\Gamma(n;\mathbb{F}),k]$. The function

$$g(Z) := (\det Y)^{\frac{1}{2}k} \, |f(Z)| \quad , \; Z \in H(n;\mathbb{F}) \; ,$$

is invariant under modular transformations. If f is a cusp form g is bounded and there exists $Z_0 \in \mathcal{F}(n;\mathbb{F})$ satisfying

$$g(Z_0) = \max \{g(Z) \; ; \; Z \in H(n;\mathbb{F})\} .$$

Proof. The invariance under modular transformations follows from (M.2) and Theorem II.1.7. Let f be a cusp form. It suffices to examine g in $F(n;\mathbb{F})$. Proposition II.3.1 and the properties of reduced matrices yield that the set

$$\{Z = X + iY \in F(n;\mathbb{F})\ ;\ y_n \leq c\}\ ,$$

where $c > 0$, is compact. Hence it suffices to show that

$$\lim_{y_n\to\infty} g(Z) = 0\ ,\ Z \in F(n;\mathbb{F})\ .$$

Given $Z = X + iY \in F(n;\mathbb{F})$ and $T > 0$ we obtain

$$\tau(T,Y) \geq \beta\ \tau(T,\mathrm{diag}\ Y) = \beta \sum_{j=1}^{n} t_j y_j$$

from Theorem I.4.12. HADAMARD'S inequality yields $(\det Y)^{\frac{1}{2}k} \leq (\det \mathrm{diag}\ Y)^{\frac{1}{2}k}$ for $k \geq 0$. If $k < 0$, one has $(\det Y)^{\frac{1}{2}k} \leq \alpha^{-\frac{1}{2}k}(\det \mathrm{diag}\ Y)^{\frac{1}{2}k}$ in view of Theorem I.4.12. Given $\rho > 0$ the function $y \longmapsto y^{\frac{1}{2}k}\ e^{-\rho y}$ is bounded in the interval $[\frac{1}{2}\sqrt{3};\infty)$. Hence there exists $C > 0$ such that

$$(\det Y)^{\frac{1}{2}k}\ e^{-2\pi\tau(T,Y)} \leq C\ e^{-2\pi\beta\tau(T,\mathrm{diag}\ Y)}$$

for all $T > 0$. Considering the FOURIER-expansion of f according to Lemma 2.3 we observe that

$$g(Z) \leq C \sum_{T\in \mathrm{Sym}^{\tau}(n;\mathcal{O}),T>0} |\alpha(T)|\ e^{-2\pi\beta\tau(T,\mathrm{diag}\ Y)}\ .$$

In virtue of Corollary 1.8 we may distribute the limit $y_n \to \infty$ through the infinite sum. We obtain $t_n > 0$ from $T > 0$, hence

$$\lim_{y_n\to\infty} e^{-2\pi\beta\tau(T,\mathrm{diag}\ Y)} = 0\ .$$

□

If f is a cusp form, we are able to improve Lemma 1.9.

Corollary 2.5. Let $f \in [\Gamma(n;\mathbb{F}),k]_0$. There exists $C = C(f) > 0$ such that

$$|\alpha(T)| \leq C\ (\det T)^{\frac{1}{2}k}$$

holds for all FOURIER-coefficients $\alpha(T)$, $T \in \mathrm{Sym}^{\tau}(n;\mathcal{O})$, $T > 0$.

Proof. According to the lemma above there is $C_1 > 0$ such that $|f(Z)| \leq C_1 \, (\det Y)^{-\frac{1}{2}k}$ for all $Z = X + iY \in H(n;\mathbb{F})$. Given $T \in \mathrm{Sym}^{\tau}(n;\mathcal{O})$, $T > 0$, Theorem 1.2 says that

$$|\alpha(T)| = (\mathrm{vol}\,\mathrm{Sym}(n;\mathcal{O}))^{-1} e^{2\pi n} \left| \int_{C(n;\mathbb{F})} f(X+iT^{-1}) e^{-2\pi i \tau(T,X)} \, dX \right|$$

$$\leq e^{2\pi n} \, C_1 \, (\det T)^{\frac{1}{2}k} \, .$$

□

Looking for non-trivial modular forms we obtain a first negative answer.

Lemma 2.6. A modular form f of weight $k < 0$ vanishes identically.

Proof. In the case $n = 1$ the assertion is well-known, cf. [52], p. 88, Theorem 4. Let $n > 1$ then one has $f|\phi \equiv 0$ by induction hypothesis. Since f is a cusp form, g defined in Lemma 2.4 is bounded. Theorem 1.2 says that

$$|\alpha(T)| \, (\det Y)^{\frac{1}{2}k} \, e^{-2\pi\tau(T,Y)} \, , \quad T \in \mathrm{Sym}^{\tau}(n;\mathcal{O}) \, , \quad T > 0 \, ,$$

is bounded and that the bound does not depend on $Y \in \mathrm{Pos}(n;\mathbb{F})$. Hence

$$|\alpha(T)| \, \varepsilon^{\frac{1}{2}kn} \, e^{-2\pi\varepsilon \, \mathrm{tr}\, T}$$

remains bounded for $\varepsilon > 0$, thus $\alpha(T) = 0$. Therefore f vanishes identically.

□

As a consequence we derive an assertion on the existence of zeros.

Corollary 2.7. Let $f \in [\Gamma(n;\mathbb{F}),k]$, $k > 0$ and $n > 1$. Then there exists $Z \in H(n;\mathbb{F})$ such that $f(Z) = 0$.

Proof. Suppose that $f(Z) \neq 0$ for all $Z \in H(n;\mathbb{F})$. Consider

$$g : H(n;\mathbb{F}) \longrightarrow \mathbb{C} \, , \quad g(Z) := f(Z)^{-1} \, ,$$

hence $g \in [\Gamma(n;\mathbb{F}),-k]$ in view of $n > 1$. Lemma 2.6 yields a contradiction.

□

In the case $n = 1$ one can, of course, find modular forms without zeros, cf. [52], p. 84.

Now we are going to estimate the dimension of the space of modular forms. The proof is based on a clever application of the maximum principle for holomorphic functions of one variable. But first we remember the definition

$$s(n;\mathbb{F}) := \sup\{\mathrm{tr}(Y^{-1});\ X+iY \in F(n;\mathbb{F})\} .$$

Theorem 2.8. *Suppose that* $f \in [\Gamma(n;\mathbb{F}),k]$, $k \geq 0$ *possesses the* FOURIER-*expanison*

$$f(Z) = \sum_{T\in \mathrm{Sym}^{\tau}(n;\mathcal{O}),T\geq 0} \alpha(T)\, e^{2\pi i\tau(T,Z)} , \quad Z \in H(n;\mathbb{F}) .$$

If $\alpha(T) = 0$ *holds for all* $T \in \mathrm{Sym}^{\tau}(n;\mathcal{O})$, $T \geq 0$, *such that*

$$\mathrm{tr}(T) \leq \frac{k}{4\pi}\, s(n;\mathbb{F}) ,$$

then f *vanishes identically.*

Proof. The assertion is proved by induction on n. The case $n = 1$ is well-known, cf. [52], p.85, Theorem 3, if we regard $s(1;\mathbb{F}) = \frac{2}{3}\sqrt{3}$ and

$$\left[\frac{k}{4\pi}\, s(1;\mathbb{F})\right] + 1 > \frac{k}{12} .$$

Let $n > 1$ and f be a modular form satisfying the conditions above. It follows that

$$f|\phi(Z_1) = \sum_{T_1} \beta(T_1)\, e^{2\pi i\tau(T_1,Z_1)} , \quad Z_1 \in H(n-1;\mathbb{F}) ,$$

where the sum is taken over $T_1 \in \mathrm{Sym}^{\tau}(n-1;\mathcal{O})$, $T_1 \geq 0$, and $\beta(T_1) = \alpha([T_1,0])$. We obtain $s(n-1;\mathbb{F}) \leq s(n;\mathbb{F})$ from Corollary II.3.12, hence $\beta(T_1) = 0$ for all $T_1 \in \mathrm{Sym}^{\tau}(n-1;\mathcal{O})$ satisfying $T_1 \geq 0$ and $\mathrm{tr}(T_1) \leq \frac{k}{4\pi}\, s(n-1;\mathbb{F})$. Because of $f|\phi \in [\Gamma(n-1;\mathbb{F}),k]$ the induction hypothesis yields $f|\phi \equiv 0$. Thus f is a cusp form and possesses a FOURIER-expansion of the form

$$f(Z) = \sum_{T\in \mathrm{Sym}^{\tau}(n;\mathcal{O}),T>0} \alpha(T)\, e^{2\pi i\tau(T,Z)} , \quad Z \in H(n;\mathbb{F}) ,$$

where the sum is only taken over all T satisfying $\mathrm{tr}(T) > \frac{k}{4\pi}\, s(n;\mathbb{F})$. Since f is a cusp form, the function $g(Z) := (\det Y)^{\frac{1}{2}k}\, |f(Z)|$ achieves its maximum at some point $Z_0 = X_0 + iY_0 \in F(n;\mathbb{F})$ according to Lemma 2.4. Let $z = x+iy$ denote a complex variable. There exists

$\varepsilon > 0$ such that

$$Z = Z_o + zI \in H(n;\mathbb{F}) \quad \text{for} \quad y \geq -\varepsilon \ .$$

Putting $w := e^{2\pi i z}$ we define for all w such that $0 < |w| = e^{-2\pi y} \leq \rho$, $\rho = e^{2\pi\varepsilon} > 1$:

$$\varphi(w) := f(Z)\, e^{-i\lambda\, \mathrm{tr}\, Z} \quad , \quad \lambda := \frac{2\pi}{n}\left(1 + \left[\frac{k}{4\pi}\, s(n;\mathbb{F})\right]\right) .$$

Given w satisfying $0 < |w| \leq \rho$ there exists a representation

$$\varphi(w) = \sum_{T>0} \alpha(T)\, e^{2\pi i \tau(T,Z_o) - i\lambda\, \mathrm{tr}(Z_o)}\, w^{\mathrm{tr}(T) - \frac{\lambda n}{2\pi}} ,$$

where the summation is extended only over those $T \in \mathrm{Sym}^{\tau}(n;\mathcal{O})$ satisfying $\mathrm{tr}(T) > \frac{k}{4\pi}\, s(n;\mathbb{F})$, hence

$$\mathrm{tr}(T) - \frac{\lambda n}{2\pi} \geq 0$$

by definition of λ . Thus φ turns out to be a holomorphic function in the circle $\{w \in \mathbb{C}\ ;\ |w| < \rho\}$. Consider ρ_1, $\rho > \rho_1 > 1$. In view of the maximum principle we can find w_1 such that $|w_1| = \rho_1$ and $|\varphi(w_1)| \geq \varphi(1)$. One has

$$|\varphi(w)| = g(Z)\ (\det Y)^{-\frac{1}{2}k}\ e^{\lambda\, \mathrm{tr}(Y)} .$$

From $w_1 = e^{2\pi i z_1}$, $z_1 = x_1 + iy_1$ and $|w_1| = e^{-2\pi y_1} = \rho_1$ we obtain

$$|\varphi(1)| = g(Z_o)\ (\det Y_o)^{-\frac{1}{2}k}\ e^{\lambda\, \mathrm{tr}(Y_o)}$$

$$\leq |\varphi(w_1)| = g(Z_1)\ (\det Y_1)^{-\frac{1}{2}k}\ e^{\lambda\, \mathrm{tr}(Y_o) + n\lambda y_1} .$$

Now $g(Z_1) \leq g(Z_o)$ yields $g(Z_o) \leq g(Z_o)\, e^{\psi(y_1)}$, where

$$\begin{aligned}\psi(y_1) &= -\tfrac{1}{2}k \log(\det Y_1) + \tfrac{1}{2}k \log(\det Y_o) + n\lambda y_1 \\ &= n\lambda y_1 - \tfrac{1}{2}k \log\left(\det(I + y_1 Y_o^{-1})\right) .\end{aligned}$$

We compute $\psi(0) = 0$ and $\psi'(0) = n\lambda - \frac{1}{2}k\, \mathrm{tr}(Y_o^{-1})$. But $Y_o \in F(n;\mathbb{F})$ leads to $\mathrm{tr}(Y_o^{-1}) \leq s(n;\mathbb{F})$ and $\psi'(0) > 0$ by definition of λ . Therefore we have $\psi(y_1) < 0$ for all y_1 satisfying $-\delta \leq y_1 < 0$ and $\delta > 0$ sufficiently small. In view of $y_1 = -\frac{1}{2\pi} \log \rho_1$ one has $\psi(y_1) < 0$ if ρ_1 is chosen sufficiently close to 1 . Thus it follows that $g(Z_o) = 0$ and $f \equiv 0$.

□

Corollary 2.9. The modular forms of weight 0 coincide with the constant functions.

Corollary 2.10. There exists $C_n > 0$ depending only on n such that

$$\dim [\Gamma(n;\mathbb{F}),k] \le C_n \cdot k^h$$

holds for all $k > 0$, where $h = n + \frac{1}{2}rn(n-1) = \dim \mathrm{Sym}(n;\mathbb{F})$.

Proof. Given $t \in \mathbb{N}$ there are at most $(4t+1)^h$ positive semi-definite matrices $T \in \mathrm{Sym}^\tau(n;\mathcal{O})$ satisfying $\mathrm{tr}\, T \le t$. Theorem 2.8 completes the proof.

□

Of course, one can choose $C_1 = \frac{1}{4}$, since

$$\dim [\Gamma(1;\mathbb{R}),k] = \begin{cases} \left[\frac{k}{12}\right] & , \text{ if } k \equiv 2 \bmod 12 , \ k \ge 0 \\ \left[\frac{k}{12}\right] + 1 & , \text{ if } k \equiv 0,4,6,8 \text{ or } 10 \bmod 12 , \\ & \quad k \ge 0 , \end{cases}$$

cf. [52], p.88, Corollary 1.

Using the estimations of $s(2;\mathbb{F})$ quoted in II,§3 we can formulate the case $n = 2$ in greater detail. The following schedule gives upper bounds for $\dim [\Gamma(2;\mathbb{F}),k]$ and "small" values of k:

k	2	4	6	8
$[\Gamma(1;\mathbb{R}),k]$	0	1	1	1
$[\Gamma(2;\mathbb{R}),k]$	0	1	1	1
$[\Gamma(2;\mathbb{C}),k]$	0	1	1	2
$[\Gamma(2;\mathbb{H}),k]$	0	1	1	1

This schedule only deals with upper bounds. The existence of a modular form of weight 4 for example will be shown in the next chapter.

But the preceding corollary also enables us to estimate the number of algebraically independent modular forms.

Definition. A set of modular forms $\{f_j \in [\Gamma(n;\mathbb{F}),k_j]; k_j \ge 0,\ 1 \le j \le m\}$ is as usual called algebraically dependent if there exists $q \in \mathbb{N}$ such that the monomials of degree q

$$f_1^{l_1}\cdot\ldots\cdot f_m^{l_m}\ , \text{ where } l_j \in \mathbb{N}_0\ , \ k_1l_1+\ldots+k_ml_m = q\ ,$$

are linearly dependent, we also say that $f_1,\ldots,f_m$ obey an <u>isobaric algebraic equation of weight</u> q . Otherwise , $f_1,\ldots,f_m$ are called <u>algebraically independent</u>.

By induction one easily proves

<u>Proposition 2.11.</u> <u>Let</u> $m,k_1,\ldots,k_m \in \mathbb{N}$, $k = \mathrm{lcm}(k_1,\ldots,k_m)$ <u>and</u> $q \in \mathbb{N}$ <u>such that</u> $q = kt$, $t \in \mathbb{N}$. <u>Then the number of</u> $(l_1,\ldots,l_m) \in \mathbb{N}_0^m$ <u>satisfying</u> $k_1l_1+\ldots+k_ml_m = q$ <u>does not remain under</u>

$$\frac{1}{(m-1)!}\, t^{m-1}\ .$$

As always let $h = n + \frac{1}{2}rn(n-1)$.

<u>Theorem 2.12.</u> <u>There is a constant</u> $c_n \in \mathbb{N}$ <u>depending only on</u> n <u>such that every</u> $h+2$ <u>modular forms</u> $f_j \in [\Gamma(n;\mathbb{F}),k_j]$, $k_j > 0$, <u>obey an isobaric algebraic equation of weight</u> $c_n(k_1\cdot\ldots\cdot k_{h+2})^{h+1}$. <u>Especially every</u> $h+2$ <u>modular forms are algebraically dependent</u>.

<u>Proof.</u> Let $m := h+2$. The monomials $f_1^{l_1}\cdot\ldots\cdot f_m^{l_m}$, where $l_j \in \mathbb{N}_0$ and $k_1l_1+\ldots+k_ml_m = q$, are modular forms of weight q . If q is a multiple of $k_1\cdot\ldots\cdot k_m$ the number of these monomials does not remain under

$$\frac{1}{(m-1)!}\ \frac{q^{m-1}}{(k_1\cdot\ldots\cdot k_m)^{m-1}} = \frac{1}{(h+1)!}\ \frac{q^{h+1}}{(k_1\cdot\ldots\cdot k_m)^{h+1}}\ .$$

Choose $c_n > (h+1)!\,C_n$, where C_n was determined in Corollary 2.10, and $q = c_n(k_1\ \ldots\ k_m)^{h+1}$. Hence there are more than C_nq^h monomials of weight q . Corollary 2.10 yields that $f_1,\ldots,f_m$ are algebraically dependent. □

If k is sufficiently large and satisfies $k \equiv 0 \bmod s(\mathbb{F})$, we shall construct $h+1$ algebraically independent modular forms of weight k (cf. Theorem V.5.8).

We shall only construct non-trivial examples of modular forms in the case $k \equiv 0 \bmod s(\mathbb{F})$. But what about the case $k \not\equiv 0 \bmod s(\mathbb{F})$? Then it is easy to derive some necessary conditions for the existence of non-trivial modular forms.

__Lemma 2.13.__ _Let_ $\mathbb{F} = \mathbb{R}, \mathbb{C}$ _and_ $k \not\equiv 0 \bmod s(\mathbb{F})$. _Then_

$$[\Gamma(n;\mathbb{F}),k] = [\Gamma(n;\mathbb{F}),k]_0 .$$

Proof. Let $f \in [\Gamma(n;\mathbb{F}),k]$, $f \not\equiv 0$. Thus we have $nk \equiv 0 \bmod s(\mathbb{F})$ as a necessary condition. But $k \not\equiv 0 \bmod s(\mathbb{F})$ yields $(n-1)k \not\equiv 0 \bmod s(\mathbb{F})$, hence $[\Gamma(n-1;\mathbb{F}),k] = \{0\}$ and $f|\phi \equiv 0$.

□

Moreover, it follows from Proposition 1.10 that

$$f([z,Z_1]) = f([Z_1,z]) = 0 \quad \text{for} \quad z \in H(1;\mathbb{F}),\ Z_1 \in H(n-1;\mathbb{F}),$$

whenever $f \in [\Gamma(n;\mathbb{F}),k]$, $k \not\equiv 0 \bmod s(\mathbb{F})$. Especially, f vanishes on all diagonal matrices in $H(n;\mathbb{F})$.

§3 Modular forms with respect to congruence subgroups

The theory of SIEGEL's modular forms with respect to congruence subgroups was developed by KOECHER [35]. Analogous results on Hermitian modular forms are due to BRAUN [8]. Here we use the same method in order to derive the corresponding assertions from §2.

As a general assumption (A) let the following hold throughout this paragraph: C denotes a congruence subgroup of level q , i.e. a group C satisfying $\Gamma(n;\mathbb{F})[q] \subset C \subset \Gamma(n;\mathbb{F})$. Let $k \in \mathbb{Z}$, $k \equiv 0 \bmod 2$ in the case $\mathbb{F} = \mathbb{H}$. ν is a <u>character</u> of C , i.e. $\nu : C \longrightarrow \{w \in \mathbb{C} \,;\, |w| = 1\}$ is a homomorphism and there exists $m = m(\nu) \in \mathbb{N}$ such that $\nu^m \equiv 1$.

We say that the assumption (A') holds if in addition $\nu(M) = 1$ for all $M \in \Gamma(n;\mathbb{F})[q]$.

A matrix $W \in \mathrm{Mat}(n;\mathbb{F})$ is called <u>rational</u>, if $\hat{W} \in \mathrm{Mat}(rn;\mathbb{Q})$. Given a rational matrix $W \in GL(n;\mathbb{F})$ let $g(W)$ denote the smallest number $g \in \mathbb{N}$ such that gW and gW^{-1} belong to $\mathrm{Mat}(n;\mathcal{O})$.

If $q \in \mathbb{N}$, put

$$GL(n;\mathcal{O})[q] := \{U \in GL(n;\mathcal{O}) \;;\; U \equiv \varepsilon I \bmod q\mathcal{O} \;,\; \varepsilon \in S(\mathbb{F})\} .$$

Clearly $GL(n;\mathcal{O})[q]$ becomes a normal subgroup of $GL(n;\mathcal{O})$ having finite index.

<u>Definition.</u> Let (A) hold. A holomorphic function $f : H(n;\mathbb{F}) \longrightarrow \mathbb{C}$ is called a <u>modular form of weight</u> k <u>on</u> $H(n;\mathbb{F})$ <u>with respect to the congruence subgroup</u> C <u>and the character</u> ν , if $f|_k M = \nu(M)f$ holds for all $M \in C$ and in the case $n = 1$ in addition $f|_k R$ is bounded in the domain $\{z \in \mathbb{C} \;;\; \mathrm{Im}\, z \geq \rho\}$, $\rho > 0$, for every $R \in SL(2;\mathbb{Q})$. Let $[C,k,\nu]$ denote the set of modular forms of weight k with respect to C and ν .

As a special case we obtain the vector space of modular forms with respect to the full modular group by

$$[\Gamma(n;\mathbb{F}),k] = [\Gamma(n;\mathbb{F}),k,1] .$$

Given $\alpha \in \mathbb{R}$, $\alpha \neq 0$, $Z \in H(n;\mathbb{F})$ and $T \in \mathrm{Sym}(n;\mathbb{F})$ we put

$$e_\alpha(T,Z) := e^{2\pi i \tau(T,Z)/\alpha}.$$

<u>Theorem 3.1.</u> <u>Let</u> (A') <u>hold</u>. <u>Given</u> $f \in [C,k,\nu]$ <u>and a rational</u> $R \in Sp(n;\mathbb{F})$ <u>there exists a</u> FOURIER-<u>expansion of the form</u>

$$f\big|_k R\,(Z) = \sum_{T \in Sym^\tau(n;\mathcal{O}),\, T \geq 0} \alpha(T;R)\; e_Q(T,Z)\;,\quad Z \in H(n;\mathbb{F})\;,$$

<u>where</u> $Q = qg(R)^2$. <u>The</u> FOURIER-<u>series converges absolutely and uniformly in every domain</u> $\{Z \in H(n;\mathbb{F})\;;\; \mathrm{Im}\, Z \geq \alpha I\}$, $\alpha > 0$. <u>The</u> FOURIER-<u>coefficients satisfy</u>:

$$\begin{aligned}
\alpha(T[U];R) &= (\det U)^{-k}\, \alpha(T;R) &&, \text{ if } \mathbb{F} = \mathbb{R},\mathbb{C}\;,\\
\alpha(T[U];R) &= \alpha(T;R) &&, \text{ if } \mathbb{F} = \mathbb{H} \text{ and } U \in GL(n;\mathcal{O})[q],\\
\alpha(T;MR) &= \nu(M)\; \alpha(T;R) && \text{for } M \in C.
\end{aligned}$$

<u>Proof.</u> Given $M \in \Gamma(n;\mathbb{F})[Q]$ one has $RMR^{-1} \in \Gamma(n;\mathbb{F})[q]$ and $(f\big|_k R)\big|_k M = (f\big|_k RMR^{-1})\big|_k R = f\big|_k R$, hence $f\big|_k R \in [\Gamma(n;\mathbb{F})[Q],k,1]$. Considering $g(Z) := f\big|_k R\,(QZ)$ we obtain a FOURIER-expansion

$$f\big|_k R\,(Z) = \sum_{T \in Sym^\tau(n;\mathcal{O})} \alpha(T;R)\; e_Q(T,Z)\;,\quad Z \in H(n;\mathbb{F})\;,$$

from Theorem 1.2. Given $U \in GL(n;\mathcal{O})[Q]$ one has $[\bar{U}',U^{-1}] \in \Gamma(n;\mathbb{F})[Q]$. Now $f\big|_k R\,(Z[U]) = (\det U)^{-k}\, f\big|_k R\,(Z)$ for $\mathbb{F} = \mathbb{R},\mathbb{C}$ resp. $f\big|_k R\,(Z[U]) = f\big|_k R\,(Z)$ for $\mathbb{F} = \mathbb{H}$ and the uniqueness of the FOURIER-coefficients lead to

$$\alpha(T[U];R) = (\det U)^{-k}\, \alpha(T;R) \quad \text{resp.} \quad \alpha(T[U];R) = \alpha(T;R)\;.$$

Finally a modification of KOECHER's principle resp. the definition, if $n = 1$, yields $\alpha(T;R) = 0$, whenever T is not positive semidefinite. Analogous arguments to Corollary 1.8 complete the proof.

□

Now we consider the vector space of modular forms.

<u>Theorem 3.2.</u> <u>Let</u> (A) <u>hold</u>.

a) $[C,k,\nu] = \{0\}$, <u>if</u> $k < 0$.

b) <u>Every modular form of weight</u> 0 <u>with respect to</u> C <u>and</u> ν <u>is constant</u>.

c) Suppose that (A') holds and $f \in [C,k,\nu]$, $k \geq 0$, possesses the FOURIER-expansion

$$f(Z) = \sum_{T \in \mathrm{Sym}^{\tau}(n;\mathcal{O}),\, T \geq 0} \alpha(T)\, e_q(T,Z) \ , \quad Z \in H(n;\mathbb{F}) \ .$$

If $i_q(n;\mathbb{F}) := [\Gamma(n;\mathbb{F}) : \Gamma(n;\mathbb{F})[q]]$ and $\alpha(T) = 0$ for all $T \in \mathrm{Sym}^{\tau}(n;\mathcal{O})$, $T \geq 0$, such that

$$\mathrm{tr}(T) \leq \frac{k}{4\pi}\, q\, i_q(n;\mathbb{F})\, s(n;\mathbb{F}) \ ,$$

then $f \equiv 0$.

Proof. Let $f \in [C,k,\nu]$. Considering f^m , $m = m(\nu)$, if $k \leq 0$, we may assume that $f \in [\Gamma(n;\mathbb{F})[q],k,1]$. Then

$$C^* := \{M \in \Gamma(n;\mathbb{F}) \ ; \ f|_k M = f\}$$

turns out to be a congruence subgroup of level q , where

$$j := [\Gamma(n;\mathbb{F}) : C^*] \leq i_q(n;\mathbb{F}) \ .$$

Let $M_1,\ldots,M_j$ denote a set of representatives of the right cosets of $\Gamma(n;\mathbb{F})$ with respect to C^* , i.e.

$$\Gamma(n;\mathbb{F}) = \bigcup_{l=1}^{j} C^* M_l \ .$$

Now define

$$g(Z) := \prod_{l=1}^{j} f|_k M_l(Z) \ ,$$

hence $g \in [\Gamma(n;\mathbb{F}),jk]$. Lemma 2.6 yields $g \equiv 0$ and therefore $f \equiv 0$ if $k < 0$. Let $k \geq 0$, then g possesses a FOURIER-expansion

$$g(Z) = \sum_{T \in \mathrm{Sym}^{\tau}(n;\mathcal{O}),\, T \geq 0} \beta(T)\, e_1(T,Z) \ , \quad Z \in H(n;\mathbb{F}) \ .$$

In the case $k = 0$ consider a FOURIER-expansion of the form in c) satisfying $\alpha(0) = 0$.

Given $k \geq 0$ it follows from the assumption and the definition of g that $\beta(T) = 0$ for all $T \in \mathrm{Sym}^{\tau}(n;\mathcal{O})$, $T \geq 0$, such that

$$\mathrm{tr}(T) \leq \frac{jk}{4\pi}\, s(n;\mathbb{F}) \ .$$

We obtain $g \equiv 0$ from Theorem 2.8, hence $f \equiv 0$.

□

The same arguments used in the proof of Corollary 2.10 lead to

Corollary 3.3. Let (A') hold. There exists a constant $C_n > 0$ depending only on n such that we have for all $k > 0$:

$$\dim\,[C,k,\upsilon] \leq C_n\ (kq\ i_q(n;\mathbb{F}))^h\ .$$

The proof of Theorem 2.12 therefore leads to

Corollary 3.4. Let (A') hold. There is a constant $c_n \in \mathbb{N}$ such that every $h+2$ modular forms $f_j \in [C,k_j,\upsilon]$, $k_j > 0$, obey an isobaric algebraic equation of weight $c_n(k_1\cdot\ldots\cdot k_{h+2})^{h+1}(q\cdot i_q(n;\mathbb{F}))^h$. Especially every $h+2$ modular forms with respect to C and υ are algebraically dependent.

The corresponding results on cusp forms with respect to congruence subgroups in the second part of [35] can be deduced in the same way.

Let (A') hold. We call $f \in [C,k,\upsilon]$ a cusp form if $f\,|_k\,R\,|\,\phi \equiv 0$ for all rational $R \in Sp(n;\mathbb{F})$. In the case $C = \Gamma(n;\mathbb{F})$ this definition is equivalent to the one given in §2. Cusp forms can also be characterized by their FOURIER-expansion, i.e. $f \in [C,k,\upsilon]$ is a cusp form if and only if it possesses a FOURIER-expansion of the form

$$f\,|_k\,R\ (Z) = \sum_{T\in Sym^T(n;\mathcal{O}),T>0} \alpha(T;R)\ e_Q(T,Z)\ ,\ Z \in H(n;\mathbb{F})\ ,$$

for all rational $R \in Sp(n;\mathbb{F})$ (cf. Theorem 3.1 and Lemma 2.3). On the other hand, a cusp form can also be defined by its growth. But proofs are omitted, since these assertions will not be used in the following.

§4 Relations between SIEGEL modular forms, Hermitian modular forms and modular forms of quaternions

SIEGEL's and Hermitian half-space form analytic submanifolds of the half-space of quaternions. If we restrict the domain of definition of modular forms of quaternions, SIEGEL's resp. Hermitian modular forms with respect to a suitable character arise.

Theorem II.1.8 says that the map $Z \longmapsto Z'$ belongs to $\mathrm{Bih}\, H(n;\mathbb{F})$, whenever $\mathbb{F} = \mathbb{C}$, $n \geq 2$, or $\mathbb{F} = \mathbb{H}$, $n = 2$. In these cases we call a modular form $f \in [C,k,\nu]$ <u>symmetric</u>, if $f(Z) = f(Z')$ holds for all $Z \in H(n;\mathbb{F})$. Let $[C,k,\nu]^s$ denote the vector space of symmetric modular forms.

It follows from Theorem II.2.3 that $\det M = 1$ for $M \in \Gamma(n;\mathbb{R})$ and $\det M \in \{\pm 1\}$ for $M \in \Gamma(n;\mathbb{C})$. Especially, det is a non-trivial character of $\Gamma(n;\mathbb{C})$. In connection with §1 we observe that

$$[\Gamma(1;\mathbb{R}),2k] = [\Gamma(1;\mathbb{C}),2k,\det^k]$$

for all $k \in \mathbb{Z}$.

One has $H(n;\mathbb{R}) \subset H(n;\mathbb{C}) \subset H(n;\mathbb{H})$. Given $f\colon \mathcal{U} \longrightarrow \mathbb{C}$ and $\mathcal{V} \subset \mathcal{U}$ let $f|_{\mathcal{V}}$ denote the restricted map $f|_{\mathcal{V}} : \mathcal{V} \longrightarrow \mathbb{C}$.

<u>Theorem 4.1.</u> <u>Let</u> $k \in \mathbb{N}_0$.

a) <u>If</u> $f \in [\Gamma(n;\mathbb{H}),2k]$, <u>then</u> $f|_{H(n;\mathbb{C})} \in [\Gamma(n;\mathbb{C}),2k,\det^k]^s$.

b) <u>If</u> $f \in [\Gamma(n;\mathbb{C}),k]$ <u>or</u> $f \in [\Gamma(n;\mathbb{C}),k,\det]$, <u>then</u> $f|_{H(n;\mathbb{R})} \in [\Gamma(n;\mathbb{R}),k]$.

<u>Proof.</u> If f is a Hermitian modular form, clearly $f|_{H(n;\mathbb{R})}$ is SIEGEL's modular form of the same weight.

Let $f \in [\Gamma(n;\mathbb{H}),2k]$ and $Z \in H(n;\mathbb{C})$. We obtain $f(Z+S) = f(Z)$ for all $S \in \mathrm{Sym}(n;\mathcal{O}(\mathbb{C})) \subset \mathrm{Sym}(n;\mathcal{O}(\mathbb{H}))$. Given $U \in GL(n;\mathcal{O}(\mathbb{C}))$ one has $M := [\bar{U}',U^{-1}] \in \Gamma(n;\mathbb{C}) \subset \Gamma(n;\mathbb{H})$ and $f(Z[U]) = f(Z)$ according to Lemma 1.7 as well as $(\det M)^k (\det U)^{2k} = 1$. If we consider Z as an element of $H(n;\mathbb{H})$, then $\det \check{Z} = (\det Z)^2$ holds. Hence it follows that $f|_{H(n;\mathbb{C})} \big|_k M = (\det M)^k f|_{H(n;\mathbb{C})}$ for all M, which belong to the set of generators of $\Gamma(n;\mathbb{C})$ quoted in Theorem II.2.3. Thus $f|_{H(n;\mathbb{C})} \in [\Gamma(n;\mathbb{C}),2k,\det^k]$.

Consider $U = e_3 I \in GL(n;\mathcal{O}(\mathbb{H}))$ and $Z \in H(n;\mathbb{C}) \subset H(n;\mathbb{H})$ then $Z[U] = Z'$. Because of $f(Z') = f(Z[U]) = f(Z)$ now $f|_{H(n;\mathbb{C})}$ turns out to be symmetric.

□

It is easy to derive the FOURIER-expansion of $f|_{H(n;\mathbb{C})}$ resp. $f|_{H(n;\mathbb{R})}$ from the FOURIER-expansion of f .

Every matrix $W \in \mathrm{Mat}(n;\mathbb{F})$ possesses a representation of the form $W = \sum_{j=1}^{r} W_j e_j$, where $W_j \in \mathrm{Mat}(n;\mathbb{R})$. We consider the maps

$$\mathrm{Re} : \mathrm{Mat}(n;\mathbb{F}) \longrightarrow \mathrm{Mat}(n;\mathbb{R}) , \quad \mathrm{Re}(W) := W_1 ,$$

$$\mathrm{Co} : \mathrm{Mat}(n;\mathbb{H}) \longrightarrow \mathrm{Mat}(n;\mathbb{C}) , \quad \mathrm{Co}(W) := W_1 e_1 + W_2 e_2 .$$

Theorem 4.2. Let $f \in [\Gamma(n;\mathbb{F}),k]$, $k \equiv 0 \bmod 2$ for $\mathbb{F} = \mathbb{H}$. If f has the FOURIER-expansion

$$f(Z) = \sum_{T \in \mathrm{Sym}^{\tau}(n;\mathcal{O}),\, T \geq 0} \alpha(T)\, e^{2\pi i \tau(T,Z)} , \quad Z \in H(n;\mathbb{F}) ,$$

then $f|_{H(n;\mathbb{R})}$ possesses the FOURIER-expansion

$$f(Z) = \sum_{R \in \mathrm{Sym}^{\tau}(n;\mathbb{Z}),\, R \geq 0} \beta(R)\, e^{2\pi i \tau(R,Z)} , \quad Z \in H(n;\mathbb{R}) ,$$

where

$$\beta(R) = \sum_{\substack{T \in \mathrm{Sym}^{\tau}(n;\mathcal{O}),\, T \geq 0 \\ \mathrm{Re}(T) = R}} \alpha(T) , \quad R \in \mathrm{Sym}^{\tau}(n;\mathbb{Z}) ,\ R \geq 0 .$$

If f is a modular form of quaternions, then

$$f(Z) = \sum_{S \in \mathrm{Sym}^{\tau}(n;\mathcal{O}(\mathbb{C})),\, S \geq 0} \gamma(S)\, e^{2\pi i \tau(S,Z)} , \quad Z \in H(n;\mathbb{C}) ,$$

turns out to be the FOURIER-expansion of $f|_{H(n;\mathbb{C})}$, where

$$\gamma(S) = \sum_{\substack{T \in \mathrm{Sym}^{\tau}(n;\mathcal{O}),\, T \geq 0 \\ \mathrm{Co}(T) = S}} \alpha(T) , \quad S \in \mathrm{Sym}^{\tau}(n;\mathcal{O}(\mathbb{C})) ,\ S \geq 0 .$$

<u>Proof.</u> Given $T \in \mathrm{Sym}^{\tau}(n;\mathcal{O})$ we observe that $\tau(T,Z) = \tau(\mathrm{Re}(T),Z)$ for $Z \in H(n;\mathbb{R})$ and $\tau(T;Z) = \tau(\mathrm{Co}(T),Z)$ for $Z \in H(n;\mathbb{C})$. Note that $\mathrm{Re}(T) \geq 0$ for every positive semi-definite $T \in \mathrm{Sym}(n;\mathbb{F})$ and $\mathrm{Co}(T) \geq 0$, whenever $T \in \mathrm{Sym}(n;\mathbb{H})$, $T \geq 0$. Finally we observe that $\mathrm{Re}(\mathrm{Sym}^{\tau}(n;\mathcal{O})) \subset \mathrm{Sym}^{\tau}(n;\mathbb{Z})$ and $\mathrm{Co}(\mathrm{Sym}^{\tau}(n;\mathcal{O}(\mathbb{H}))) \subset \mathrm{Sym}^{\tau}(n;\mathcal{O}(\mathbb{C}))$ Because of the absolute convergence the FOURIER-series may be rearranged. Thus we obtain the assertion.

□

It is an interesting but unsolved problem how those modular forms, which arise by restriction of the domain of definition, can be described. A partial answer is given in the next chapter, where we consider theta-series, since the restriction of a theta-series turns out to be a theta-series again (cf. Corollary IV.2.8).

Chapter IV Theta-series

In this chapter we will give first examples of non-identically vanishing modular forms by means of theta-series. By application of the KRONECKER-product of matrices it is possible to confine the examination of theta-series on SIEGEL's or Hermitian half-space to the case "$S = (1)$". But the properties of the KRONECKER-product essentially depend on the law of commutativity in the field. Thus we cannot proceed in the same way if we consider theta-series on the half-space of quaternions. Instead of this we use methods developed by KRAZER [39], which base on a precise examination of theta-series with characteristics. The translation of KRAZER's results into modern language and notations by VOLKMANN [60] was very helpful to the author.

§1 deals with convergence and elementary properties of theta-series with characteristics. By means of a general Theta-transformation-formula we derive conditions on S in order to obtain a modular form with respect to the full modular group. Those "stable" matrices enable us to construct non-identically vanishing modular forms of any degree, whenever the weight is a multiple of 4.

In §3 the behavior of theta-series under modular transformations is examined in greater detail, whenever S represents an even quadratic form. These results are used to describe singular modular forms by theta-series. In this chapter we aim to deduce a theorem saying that every modular form turns out to be a linear combination of theta-series, if the weight is sufficiently small in proportion to the degree. On the other hand, singular modular forms can be characterized in virtue of their FOURIER-expansions.

§1 Elementary properties of theta-series

The classical theta-series

$$(1) \qquad \Theta(y) := \sum_{g\in\mathbb{Z}} e^{-\pi y g^2}, \quad y > 0,$$

converges absolutely in $\mathbb{R}^+$. A lot of generalizations have been developed in the meantime. Theta-series on $H(n;\mathbb{R})$ resp. $H(n;\mathbb{C})$ were introduced by SIEGEL [53] resp. BRAUN [7]. RESNIKOFF [50] investigated corresponding series on finitely dimensional formal real JORDAN-algebras. We consider theta-series with characteristics.

Given $A = U + iV$ and B in $\mathrm{Mat}(m,n;\mathbb{F})_{\mathbb{C}} := \mathrm{Mat}(m,n;\mathbb{F}) \otimes_{\mathbb{R}} \mathbb{C}$ we define

$$\bar{A}' := \bar{U}' + i\bar{V}', \quad \tilde{A} := \bar{U}' - i\bar{V}'$$

$$\tau(A,B) := \tfrac{1}{2}\mathrm{tr}(A\bar{B}' + B\bar{A}') \in \mathbb{C}.$$

Every $A \in \mathrm{Mat}(m,n;\mathbb{F})_{\mathbb{C}}$ can uniquely be represented in the form

$$(2) \qquad A = (a_{kl}), \quad a_{kl} = \sum_{j=1}^{r} a_{kl}^{(j)} e_j, \quad a_{kl}^{(j)} \in \mathbb{C}, \ 1 \leq j \leq r, \quad 1 \leq k \leq m, \ 1 \leq l \leq n.$$

Given A,B in the form (2) one computes

$$(3) \qquad \tau(A,B) = \sum_{k=1}^{m} \sum_{l=1}^{n} \sum_{j=1}^{r} a_{kl}^{(j)} b_{kl}^{(j)}.$$

Thus one easily checks the following elementary properties of τ.

Proposition 1.1. a) Suppose that $S = S^{(m)} = \bar{S}'$, $T = T^{(n)} = \bar{T}'$ and $A,B \in \mathrm{Mat}(m,n;\mathbb{F})_{\mathbb{C}}$. Then

$$\tau(A,B) = \tau(B,A) = \tau(\bar{A}',\bar{B}'),$$

$$\tau(S[A+B],T) = \tau(S[A] + S[B],T) + 2\tau(SA,BT).$$

b) Given $A = A^{(m,n)}$, $B = B^{(n,p)}$ and $C = C^{(m,p)}$ one has

$$\tau(AB,C) = \tau(A,C\bar{B}') = \tau(B,\bar{A}'C).$$

Now we state the formal definition of theta-series.

Definition. Suppose that $Z \in H(n;\mathbb{F})$, $S \in \mathrm{Pos}(m;\mathbb{F})$, $P,Q \in \mathrm{Mat}(m,n;\mathbb{F})_{\mathbb{C}}$ and that Λ is a lattice in $\mathrm{Mat}(m,n;\mathbb{F})$. Then

$$\Theta_{P,Q}(Z,S;\Lambda) := \sum_{G\in\Lambda} e^{\pi i\tau(S[G+P],Z) + 2\pi i\tau(Q,G+P)}$$

is called theta-series on $H(n;\mathbb{F})$ in Z and S with respect to the characteristic (P,Q) and the lattice Λ .

We call a function $f : \mathrm{Mat}(m,n;\mathbb{F})_{\mathbb{C}} \longrightarrow \mathbb{C}$, $A \longmapsto f(A)$, holomorphic if f is a holomorphic function of the coefficients $a_{kl}^{(j)}$ of the representation (2) in the usual sense.

The behavior of convergence of theta-series is stated in

Theorem 1.2. a) $\Theta_{P,Q}(Z,S;\Lambda)$ converges absolutely and locally uniformly in dependence on Z,S,P and Q .

b) Given $\rho > 0$ then $\Theta_{P,Q}(Z,S;\Lambda)$ converges absolutely and uniformly in the domain $\mathrm{Im}\, Z \geq \rho I$, $S \geq \rho I$, $P \in \mathrm{Mat}(m,n;\mathbb{F})$, $\tau(P,P) \leq \rho^{-2}$, $\tau(\mathrm{Im}\, Q,\mathrm{Im}\, Q) \leq \rho^{-2}$.

c) $\Theta_{P,Q}(Z,S;\Lambda)$ considered as a function of Z is holomorphic on the half-space $H(n;\mathbb{F})$, and considered as a function of P resp. Q the theta-series turns out to be a holomorphic function on $\mathrm{Mat}(m,n;\mathbb{F})_{\mathbb{C}}$.

Proof. The series of absolute values of $\Theta_{P,Q}(Z,S;\Lambda)$ is given by

$$\sum_{G\in\Lambda} e^{\pi\varphi(G)} ,$$

where

$$\varphi(G) = \mathrm{Re}\{i\tau(S[G+P],Z) + 2i\tau(Q,G+P)\} .$$

Putting $P = P_1 + iP_2$, $Q = Q_1 + iQ_2$ and $Z = X + iY$ Proposition 1.1 yields

$$\varphi(G) = -\tau(S[G+P_1],Y) + \tau(S[P_2],Y) - 2\tau(S(G+P_1),P_2X)$$
$$-2\tau(Q_2,G+P_1) - 2\tau(Q_1,P_2) .$$

a) Since X,Y,S,P_1,P_2,Q_1,Q_2 belong to compact subsets, there exists $\gamma > 0$ satisfying

$$\varphi(G) \leq -\tau(S[G],Y) + 2\,\tau(G,SP_1Y + SP_2X + Q_2) + \gamma$$

for all $G \in \Lambda$ in view of Proposition 1.1. It follows from the CAUCHY-SCHWARZ inequality that

$$|\tau(G,A)| \leq \tau(G,G)^{\frac{1}{2}}\,\tau(A,A)^{\frac{1}{2}}$$

holds for all $A \in Mat(m,n;\mathbb{F})$. Hence there are $\alpha > 0$ and $\beta > 0$ such that

$$\varphi(G) \leq -2\alpha\tau(G,G) + \beta\tau(G,G)^{\frac{1}{2}} + \gamma$$

for all $G \in \Lambda$. Thus all but finitely many $G \in \Lambda$ satisfy

$$\varphi(G) \leq -\alpha\tau(G,G)\ .$$

Let $G_1,\ldots,G_q$, $q = rmn$, denote a basis of Λ and let T be the GRAM-matrix of τ with respect to this basis, i.e. $T = (t_{kl}) \in Sym(q;\mathbb{R})$, $t_{kl} = \tau(G_k,G_l)$. Since τ is positive definite, one has $T \in Pos(q;\mathbb{R})$. Hence there exists $\delta > 0$ satisfying $T \geq \delta I$. Given $G = \gamma_1G_1 + \ldots + \gamma_qG_q \in \Lambda$, where $\mathfrak{g} = (\gamma_1,\ldots,\gamma_q)' \in \mathbb{Z}^q$, one calculates

$$\tau(G,G) = T[\mathfrak{g}] \geq \delta\mathfrak{g}'\mathfrak{g}\ .$$

Thus the series of absolute values is dominated by

$$\sum_{G\in\Lambda} e^{-\pi\alpha\tau(G,G)} = \sum_{\mathfrak{g}\in\mathbb{Z}^q} e^{-\pi\alpha T[\mathfrak{g}]}$$

$$\leq \sum_{\mathfrak{g}\in\mathbb{Z}^q} e^{-\pi\alpha\delta\mathfrak{g}'\mathfrak{g}} = \Theta(\alpha\delta)^q\ .$$

We obtain the absolute and locally uniform convergence of the theta-series by the convergence of (1)

b) $Y \geq \rho I$, $S \geq \rho I$ and $P_2 = 0$ yield

$$\varphi(G) = -\tau(S[G+P],Y) - 2\tau(Q_2,G+P)$$
$$\leq -\rho^2\tau(G+P,G+P) - 2\tau(Q_2,G+P)\ .$$

Now $\tau(P,P) \leq \rho^{-2}$, $\tau(Q_2,Q_2) \leq \rho^{-2}$ and the CAUCHY-SCHWARZ inequality lead to

$$\varphi(G) \leq -\rho^2\tau(G,G) + 2(\rho+\rho^{-1})\tau(G,G)^{\frac{1}{2}} + 2\rho^{-2}\ .$$

We obtain the assertion from the proof of part a).

c) The series converges locally uniformly. Hence the function is holomorphic, because each term of the series is holomorphic. □

We remember the definition of the dual lattice Λ^τ in I, §1, i.e.

$$\Lambda^\tau := \{K \in \mathrm{Mat}(m,n;\mathbb{F}) \; ; \; \tau(G,K) \in \mathbb{Z} \; \text{ for all } \; G \in \Lambda\} .$$

The special case "$S = (1)$, $\Lambda = \mathbb{Z}^n$" of the following assertion is due to KRAZER [39], I,§5.

Proposition 1.3. Let $H \in \Lambda$, $K \in \Lambda^\tau$ and $U \in \mathrm{Mat}(m,n;\mathbb{F})_\mathbb{C}$.

a) $\quad \Theta_{P+H,Q}(Z,S;\Lambda) = \Theta_{P,Q}(Z,S;\Lambda)$.

b) $\quad \Theta_{P,Q+K}(Z,S;\Lambda) = e^{2\pi i\tau(P,K)}\, \Theta_{P,Q}(Z,S;\Lambda)$.

c) $\quad \Theta_{P,Q+SUZ}(Z,S;\Lambda) = e^{-\pi i\tau(S[U],Z) - 2\pi i\tau(Q,U)}\, \Theta_{P+U,Q}(Z,S;\Lambda)$.

Proof. a) Together with G also $G+H$ runs through Λ. Because of the absolute convergence we may rearrange the series.

b) One has $\tau(K,G+P) = \tau(K,G) + \tau(K,P) \equiv \tau(K,P) \bmod 1$ for all $G \in \Lambda$.

c) Proposition 1.1 yields

$$\tau(S[G+P],Z) + 2\tau(Q+SUZ,G+P)$$
$$= \tau(S[G+P+U],Z) + 2\tau(Q,G+P+U) - \tau(S[U],Z) - 2\tau(Q,U) .$$

□

The validity of the identities in Proposition 1.3 suffices to characterize the function to a certain extent. We formulate a generalization of Satz XIII by KRAZER [39], p.34, in the following

Theorem 1.4. Let $\varphi : \mathrm{Mat}(m,n;\mathbb{F})_\mathbb{C} \longrightarrow \mathbb{C}$ be holomorphic. Suppose that Λ is a lattice in $\mathrm{Mat}(m,n;\mathbb{F})$, $Z \in H(n;\mathbb{F})$, $S \in \mathrm{Pos}(m;\mathbb{F})$ and $P,Q \in \mathrm{Mat}(m,n;\mathbb{F})_\mathbb{C}$ such that

(i) $\quad \varphi(W+K) = e^{2\pi i\tau(P,K)}\, \varphi(W)$,

(ii) $\quad \varphi(W+SHZ) = e^{-\pi i\tau(S[H],Z) - 2\pi i\tau(W+Q,H)}\, \varphi(W)$

hold for all $H \in \Lambda$, $K \in \Lambda^\tau$ and $W \in \mathrm{Mat}(m,n;\mathbb{F})_\mathbb{C}$. Then there exists $\chi \in \mathbb{C}$ satisfying

$$\varphi(W) = \chi\, \Theta_{P,Q+W}(Z,S;\Lambda)$$

for all $W \in \mathrm{Mat}(m,n;\mathbb{F})_\mathbb{C}$.

<u>Proof.</u> Put $\psi(W) := e^{-2\pi i\tau(P,W)}\varphi(W)$. Then we obtain $\psi(W+K) = \psi(W)$ for all $K \in \Lambda^\tau$ from (i). On the analogy of the proof of Theorem III. 1.2 we conclude that ψ can be developed into a FOURIER-series

$$\psi(W) = \sum_{G\in\Lambda} \alpha(G)\, e^{2\pi i\tau(G,W)} ,$$

$$\varphi(W) = \sum_{G\in\Lambda} \alpha(G)\, e^{2\pi i\tau(G+P,W)} .$$

We determine the FOURIER-coefficients. Given $H \in \Lambda$ we compute on the one hand that

$$\varphi(W+SHZ) = \sum_{G\in\Lambda} \alpha(G)\, e^{2\pi i\tau(G+P,W+SHZ)}$$

and on the other hand because of (ii) that

$$\varphi(W+SHZ) = e^{-\pi i\tau(S[H],Z) - 2\pi i\tau(W+Q,H)}\varphi(W)$$

$$= \sum_{G\in\Lambda} \alpha(G)\, e^{-\pi i\tau(S[H],Z) - 2\pi i\tau(Q,H) + 2\pi i\tau(G+P-H,W)} .$$

The uniqueness of the FOURIER-coefficients implies that

$$\alpha(G)\, e^{2\pi i\tau(G+P,SHZ)} = \alpha(G+H)\, e^{-\pi i\tau(S[H],Z) - 2\pi i\tau(Q,H)} .$$

Hence $G = 0$ leads to

$$\alpha(H) = \alpha(0)\, e^{\pi i\tau(S[H],Z) + 2\pi i\tau(P,SHZ) + 2\pi i\tau(Q,H)}$$

for all $H \in \Lambda$. Because of $\tau(S[H],Z) + 2\tau(P,SHZ) = \tau(S[H+P],Z) - \tau(S[P],Z)$ according to Proposition 1.1 summation over $H \in \Lambda$ yields

$$\varphi(W) = \alpha(0)\, e^{-\pi i\tau(S[P],Z) - 2\pi i\tau(Q,P)}\,\Theta_{P,Q+W}(Z,S;\Lambda) .$$

□

The preceding assertion turns out to be the crucial tool in order to determine the behavior of theta-series under modular transformations.

Next we describe a "symmetry" in the arguments.

<u>Proposition 1.5.</u> <u>Let</u> $Y \in \mathrm{Pos}(n;\mathbb{F})$ <u>and</u> $S \in \mathrm{Pos}(m;\mathbb{F})$ <u>then</u>

$$\Theta_{P,Q}(iY,S;\Lambda) = \Theta_{\bar{P}',\bar{Q}'}(iS,Y;\bar{\Lambda}') .$$

<u>Proof.</u> Proposition 1.1 yields $\tau(S[G+P],Y) = \tau(S,Y[\bar{G}'+\bar{P}'])$ and $\tau(Q,G+P) = \tau(\bar{Q}',\bar{G}'+\bar{P}')$. Now $\bar{\Lambda}' := \{\bar{G}' ; G \in \Lambda\}$ completes the proof. □

Now we state connections between theta-series with respect to different lattices.

Proposition 1.6. *Let* Λ, Λ^* *be two lattices in* $Mat(m,n;\mathbb{F})$ *and suppose that* $U \in GL(n;\mathbb{F})$, $V \in GL(m;\mathbb{F})$ *such that the map*

$$\Lambda \longrightarrow \Lambda^* , \quad G \longmapsto VGU ,$$

is bijective. Then one has

$$\Theta_{P,Q}(Z,S;\Lambda^*) = \Theta_{V^{-1}PU^{-1},\bar{V}'Q\bar{U}'}(Z[\bar{U}'],S[V];\Lambda) .$$

Proof. Given $G \in \Lambda$ Proposition 1.1 yields

$$\tau(S[V][G+V^{-1}PU^{-1}],Z[\bar{U}']) = \tau(S[VGU+P],Z) ,$$

$$\tau(\bar{V}'Q\bar{U}',G+V^{-1}PU^{-1}) = \tau(Q,VGU+P) .$$

□

In addition, consider the special case

$$\Lambda(\mathbb{F}) := Mat(m,n;\mathcal{O}(\mathbb{F})) .$$

Then $\Lambda(\mathbb{F})^{\tau} = Mat(m,n;\mathcal{O}^{\#})$ and Lemma I.1.15 leads to

Corollary 1.7. a) *Given* $U \in GL(n;\mathcal{O})$ *and* $V \in GL(n;\mathcal{O})$ *then*

$$\Theta_{P,Q}(Z,S;\Lambda(\mathbb{F})) = \Theta_{V^{-1}P\bar{U}'^{-1},\bar{V}'QU}(Z[U],S[V];\Lambda(\mathbb{F})) .$$

b) *Put* $\alpha = e_1 + e_2$ *then*

$$\Theta_{P,Q}(Z,S;\Lambda(\mathbb{H})^{\tau}) = \Theta_{\alpha^{-1}P,\bar{\alpha}Q}(Z,S[\alpha I];\Lambda(\mathbb{H}))$$

$$= \Theta_{P\bar{\alpha}^{-1},Q\alpha}(Z[\alpha I],S;\Lambda(\mathbb{H})) .$$

What does happen if we restrict a theta-series onto Hermitian or SIEGEL's half-space? The arising function turns out to be a theta-series, too.

We remember the definitions

$$\hat{} : Mat(m,n;\mathbb{F})_{\mathbb{C}} \longrightarrow Mat(rm,rn;\mathbb{R})_{\mathbb{C}} , \ W = W_1 + iW_2 \longmapsto \hat{W} = \hat{W}_1 + i\hat{W}_2 ,$$

$$\check{} : Mat(m,n;\mathbb{H})_{\mathbb{C}} \longrightarrow Mat(2m,2n;\mathbb{C})_{\mathbb{C}} , \ W = W_1 + iW_2 \longmapsto \check{W} = \check{W}_1 + i\check{W}_2 .$$

Clearly, the maps

$$\nu: \mathrm{Mat}(m,n;\mathbb{F})_{\mathbb{C}} \to \mathrm{Mat}(rm,n;\mathbb{R})_{\mathbb{C}}, \ A = (a_1,\dots,a_n) \mapsto (\mathfrak{b}_1,\dots,\mathfrak{b}_n),$$

$$\mu: \mathrm{Mat}(m,n;\mathbb{H})_{\mathbb{C}} \to \mathrm{Mat}(2m,n;\mathbb{C})_{\mathbb{C}}, \ A = (a_1,\dots,a_n) \mapsto (\mathfrak{c}_1,\dots,\mathfrak{c}_n),$$

whenever $\mathfrak{b}_j$ resp. $\mathfrak{c}_j$ denotes the first column of $\hat{a}_j \in \mathrm{Mat}(rm,r;\mathbb{R})_{\mathbb{C}}$ resp. of $\check{a}_j \in \mathrm{Mat}(2m,2;\mathbb{C})_{\mathbb{C}}$, become isomorphisms of the vector spaces. Given a lattice Λ in $\mathrm{Mat}(m,n;\mathbb{F})$ then $\nu(\Lambda)$ resp. $\mu(\Lambda)$ for $\mathbb{F}=\mathbb{H}$ turns out to be a lattice, too.

Let $A,B \in \mathrm{Mat}(m,n;\mathbb{F})_{\mathbb{C}}$, then one computes that

$$(4) \qquad \begin{cases} \tau(A,B) = \tau(\nu(A),\nu(B)), \\ \tau(A,B) = \tau(\mu(A),\mu(B)), \text{ if } \mathbb{F} = \mathbb{H}. \end{cases}$$

Remember the definitions of the maps Re, Co in III, §4. Hence given $S \in \mathrm{Sym}(m;\mathbb{F})$ and $A \in \mathrm{Mat}(m,n;\mathbb{F})$ one easify verifies

$$(5) \qquad \begin{cases} \mathrm{Re}(S[A]) = \hat{S}[\nu(A)], \\ \mathrm{Co}(S[A]) = \check{S}[\mu(A)], \text{ if } \mathbb{F} = \mathbb{H}. \end{cases}$$

Given $Z \in H(n;\mathbb{R}) \subset H(n;\mathbb{F})$ and $T \in \mathrm{Sym}(n;\mathbb{F})$ one has

$$(6) \qquad \tau(T,Z) = \tau(\mathrm{Re}(T),Z)$$

and, in addition, if $Z \in H(n;\mathbb{C}) \subset H(n;\mathbb{H})$ and $T \in \mathrm{Sym}(n;\mathbb{H})$

$$(7) \qquad \tau(T,Z) = \tau(\mathrm{Co}(T),Z).$$

A straightforward computation using (4),(5),(6) and (7) leads to

<u>Lemma 1.8.</u> <u>Let</u> Λ <u>be a lattice in</u> $\mathrm{Mat}(m,n;\mathbb{F})$, $S \in \mathrm{Pos}(m;\mathbb{F})$ <u>and</u> $P,Q \in \mathrm{Mat}(m,n;\mathbb{F})_{\mathbb{C}}$. <u>Then</u>

$$\Theta_{P,Q}(\cdot,S;\Lambda)\Big|_{H(n;\mathbb{R})} = \Theta_{\nu(P),\nu(Q)}(\cdot,\hat{S};\nu(\Lambda))$$

<u>holds</u>. <u>If</u> $\mathbb{F} = \mathbb{H}$ <u>one has in addition</u>

$$\Theta_{P,Q}(\cdot,S;\Lambda)\Big|_{H(n;\mathbb{C})} = \Theta_{\mu(P),\mu(Q)}(\cdot,\check{S};\mu(\Lambda)).$$

What about the special case $\Lambda = \Lambda(\mathbb{F})$? Clearly

$$\nu(\Lambda(\mathbb{C})) = \mathrm{Mat}(2m,n;\mathbb{Z}).$$

Now put

$$u = \begin{pmatrix} \frac{1}{2} & 1 & 0 & 0 \\ \frac{1}{2} & 0 & 1 & 0 \\ \frac{1}{2} & 0 & 0 & 1 \\ \frac{1}{2} & 0 & 0 & 0 \end{pmatrix} \in GL(4;\mathbb{R}) \ ,$$

$$v = \begin{pmatrix} \frac{1}{2}(e_1 + e_2) & 0 \\ \frac{1}{2}(e_1 + e_2) & 1 \end{pmatrix} \in GL(2;\mathbb{C}) \ .$$

and $U = [u,\ldots,u] \in GL(4m;\mathbb{R})$, $V = [v,\ldots,v] \in GL(2m;\mathbb{C})$. A verification yields

$$\nu(\Lambda(\mathbb{H})) = \{UG \ ; \ G \in Mat(4m,n;\mathbb{Z})\} \ ,$$

$$\mu(\Lambda(\mathbb{H})) = \{VG \ ; \ G \in Mat(2m,n;\mathcal{O}(\mathbb{C}))\} \ .$$

Hence Lemma 1.8 and Proposition 1.6 lead to

<u>Corollary 1.9.</u> <u>Let</u> $S \in Pos(m;\mathbb{F})$, $P,Q \in Mat(m,n;\mathbb{F})_{\mathbb{C}}$. <u>Then</u>

$$\Theta_{P,Q}(\cdot,S;\Lambda(\mathbb{C}))\Big|_{H(n;\mathbb{R})} = \Theta_{\nu(P),\nu(Q)}(\cdot,\hat{S};\Lambda(\mathbb{R})) \ ,$$

$$\Theta_{P,Q}(\cdot,S;\Lambda(\mathbb{H}))\Big|_{H(n;\mathbb{R})} = \Theta_{U^{-1}\nu(P),U'\nu(Q)}(\cdot,\hat{S}[U];\Lambda(\mathbb{R})) \ ,$$

$$\Theta_{P,Q}(\cdot,S;\Lambda(\mathbb{H}))\Big|_{H(n;\mathbb{C})} = \Theta_{V^{-1}\mu(P),\bar{V}'\mu(Q)}(\cdot,\check{S}[V];\Lambda(\mathbb{C})) \ .$$

Next we consider theta-series, whenever S belongs to a special class.

<u>Definition.</u> A matrix $S \in Sym(m;\mathbb{F})$ is called <u>even</u> or, more precisely, $\mathbb{F}$-<u>even</u> if $\frac{1}{2}S$ belongs to $Sym^{\tau}(m;\mathcal{O}(\mathbb{F}))$.

Lemma III.1.1 is reformulated in the following

<u>Proposition 1.10.</u> <u>Given</u> $S \in Sym(m;\mathbb{F})$ <u>the following assertions are equivalent</u>:

(i) S <u>is</u> $\mathbb{F}$-<u>even</u>.

(ii) $\tau(S,T) \in 2\mathbb{Z}$ <u>for all</u> $T \in Sym(m;\mathcal{O})$.

(iii) $S[\mathfrak{g}] \in 2\mathbb{Z}$ <u>for all</u> $\mathfrak{g} \in \mathcal{O}^m$.

(iv) $s_j \in 2\mathbb{Z}$, $1 \leq j \leq m$, $s_{jk} \in \mathcal{O}^{\#}$, $1 \leq j < k \leq m$.

If S *is* $\mathbb{F}$-*even and* $G \in Mat(m,n;\mathcal{O})$ *then* $S[G]$ *becomes* $\mathbb{F}$-*even, too.*

Now consider theta-series in even matrices S .

Proposition 1.11. *Let* $S \in Pos(m;\mathbb{F})$ *be* $\mathbb{F}$-*even. Then*

$$\Theta_{P,Q}(Z+T,S;\Lambda(\mathbb{F})) = e^{-\pi i\tau(S[P],T)}\,\Theta_{P,Q+SPT}(Z,S;\Lambda(\mathbb{F}))$$

holds for all $Z \in H(n;\mathbb{F})$, $T \in Sym(n;\mathcal{O})$ *and* $P,Q \in Mat(m,n;\mathbb{F})_{\mathbb{C}}$.

Proof. Given $G \in \Lambda(\mathbb{F})$ Proposition 1.1 leads to

$$\begin{aligned}&\tau(S[G+P],T) + 2\tau(Q,G+P)\\ &= \tau(S[G],T) - \tau(S[P],T) + 2\tau(Q+SPT,G+P)\\ &\equiv -\tau(S[P],T) + 2\tau(Q+SPT,G+P) \mod 2 ,\end{aligned}$$

since $\tau(S[G],T) \in 2\mathbb{Z}$ holds in view of the preceding proposition.

□

Remember the definitions of the matrices U resp. V above.

Proposition 1.12. a) *Let* $S \in Sym(m;\mathbb{H})$ *be even. Then* $\check{S}[V]$ *becomes* $\mathbb{C}$-*even and* $\hat{S}[U]$ *becomes* $\mathbb{R}$-*even.*

b) *If* $S \in Sym(m;\mathbb{C})$ *is even, then* $\hat{S}$ *is* $\mathbb{R}$-*even, too.*

Proof. Given $\mathfrak{g} \in \mathcal{O}(\mathbb{C})^{2m}$ the choice of V leads to the existence of a uniquely determined $\mathfrak{h} \in \mathcal{O}(\mathbb{H})^m$ such that $V\mathfrak{g}$ becomes the first column of $\check{\mathfrak{h}}$. Because of $\check{S}[V][\mathfrak{g}] = S[\mathfrak{h}]$ Proposition 1.10 yields the assertion. The other statements are proved in an analogous way.

□

In view of symmetric modular forms a verification yields

Proposition 1.13. *Let* $Z \in H(n;\mathbb{C})$, $S \in Pos(m;\mathbb{C})$ *and* $P,Q \in Mat(m,n;\mathbb{C})$. *Then*

$$\Theta_{P,Q}(Z',S;\Lambda(\mathbb{C})) = \Theta_{\bar{P},\bar{Q}}(Z,S';\Lambda(\mathbb{C})) .$$

Finally, we examine what happens if Z or S decomposes into blocks.

<u>Lemma 1.14.</u> <u>On corresponding assumptions on the numbers of rows and columns one has for</u> $P = (P_1, P_2)$ <u>and</u> $Q = (Q_1, Q_2)$:

$$\Theta_{P,Q}([Z_1, Z_2], S; \Lambda(\mathbb{F})) = \Theta_{P_1,Q_1}(Z_1, S; \Lambda(\mathbb{F}))\ \Theta_{P_2,Q_2}(Z_2, S; \Lambda(\mathbb{F})) .$$

<u>and for</u> $P = \begin{pmatrix} P_1 \\ P_2 \end{pmatrix}$, $Q = \begin{pmatrix} Q_1 \\ Q_2 \end{pmatrix}$:

$$\Theta_{P,Q}(Z, [S_1, S_2]; \Lambda(\mathbb{F})) = \Theta_{P_1,Q_1}(Z, S_1; \Lambda(\mathbb{F}))\ \Theta_{P_2,Q_2}(Z, S_2; \Lambda(\mathbb{F})) .$$

<u>Proof.</u> Decomposing $G = (G_1, G_2)$ one calculates for $Z = [Z_1, Z_2]$:

$$\tau(S[G+P], Z) = \tau(S[G_1+P_1], Z_1) + \tau(S[G_2+P_2], Z_2) ,$$

$$\tau(Q, G+P) = \tau(Q_1, G_1+P_1) + \tau(Q_2, G_2+P_2) .$$

Because of the absolute convergence the assertion follows by multiplication of the series on the right side. The second identity is proved in the same way.

□

In the case $P = Q = 0$ one has

$$\Theta_{0,0}([Z_1, Z_2], S; \Lambda(\mathbb{F})) = \Theta_{0,0}(Z_1, S; \Lambda(\mathbb{F}))\ \Theta_{0,0}(Z_2, S; \Lambda(\mathbb{F})) .$$

Thus considering theta-series "without characteristics" Proposition III.1.10 can be stated in this simple form.

§2 Theta-series as modular forms

In this paragraph we construct modular forms by means of theta-series. The representation leans on KRAZER [39],III, and KOECHER [37],II, §4,§6 , where the KRONECKER-product will not be used.

Since the half-space $H(n;\mathbb{F})$ is convex, we obtain a holomorphic function $\varphi : H(n;\mathbb{F}) \longrightarrow \mathbb{C}$ satisfying $\varphi(Z)^2 = \det(\frac{1}{i}Z)$ and $\varphi(iY) = \sqrt{\det Y}$. This function is uniquely determined in view of Lemma II.1.6 and is denoted by $(\det \frac{1}{i}Z)^{\frac{1}{2}}$.

Lemma 2.1. Given $Z \in H(n;\mathbb{F})$, $S \in \mathrm{Pos}(m;\mathbb{F})$ and $W \in \mathrm{Mat}(m,n;\mathbb{F})_{\mathbb{C}}$ one has

$$\int_{\mathrm{Mat}(m,n;\mathbb{F})} e^{\pi i \tau(S[U+W],Z)}\, dU = (\det S)^{-\frac{1}{2}rn} (\det \tfrac{1}{i}Z)^{-\frac{1}{2}rm}$$

Proof. Since the integrand possesses an integrable majorant, whenever Z belongs to a compact subset, both sides of the equation represent holomorphic functions. In view of Lemma II.1.6 it suffices to consider $Z = iY$, $Y \in \mathrm{Pos}(n;\mathbb{F})$. According to Theorem I.3.6 there exist $F \in GL(n;\mathbb{F})$ and $G \in GL(m;\mathbb{F})$ such that $Y = \bar{F}'F$ and $S = \bar{G}'G$. Putting $V := GW\bar{F}'$ one has

$$\tau(S[U+W],Y) = \tau(GU\bar{F}'+V, GU\bar{F}'+V) \quad .$$

On the analogy of [38],III.4.8, one proves that the absolute value of the JACOBIan determinant of the map

$$\mathrm{Mat}(m,n;\mathbb{F}) \longrightarrow \mathrm{Mat}(m,n;\mathbb{F}) \;, \quad U \longmapsto GU\bar{F}' \;,$$

is given by $|\det \hat{G}|^n \, |\det \hat{F}|^m = (\det S)^{\frac{1}{2}rn} (\det Y)^{\frac{1}{2}rm}$. Change of variables yields

$$\int_{\mathrm{Mat}(m,n;\mathbb{F})} e^{-\pi\tau(GU\bar{F}'+V, GU\bar{F}'+V)}\, dU$$

$$= (\det S)^{-\frac{1}{2}rn} (\det Y)^{-\frac{1}{2}rm} \int_{\mathrm{Mat}(m,n;\mathbb{F})} e^{-\pi\tau(U+V,U+V)}\, dU \;.$$

Since the last integral consists of rmn integrals of the form

$$\int_{-\infty}^{+\infty} e^{-\pi(u+v)^2}\, du \;, \quad v \in \mathbb{C} \;,$$

each of which equals 1 , the proof is complete. □

If L is a lattice in a Euclidean vector space let $\operatorname{vol} L$ denote the Euclidean volume of a fundamental parallelotope of L.

<u>Theorem 2.2.</u> <u>Theta-transformation-formula</u>
Let $Z \in H(n;\mathbb{F})$, $S \in \operatorname{Pos}(m;\mathbb{F})$, $P,Q \in \operatorname{Mat}(m,n;\mathbb{F})_{\mathbb{C}}$ *and* Λ *be a lattice in* $\operatorname{Mat}(m,n;\mathbb{F})$. *Then one has*

$$\Theta_{-Q,P}(-Z^{-1},S^{-1};\Lambda^{\tau})$$
$$= (\operatorname{vol}\Lambda)\ (\det \tfrac{1}{i}Z)^{\frac{1}{2}rm}\ (\det S)^{\frac{1}{2}rn}\ e^{-2\pi i\tau(Q,P)}\ \Theta_{P,Q}(Z,S;\Lambda)\ .$$

<u>Proof.</u> In view of Lemma II.1.6 it suffices to consider $Z = iY$, $Y \in \operatorname{Pos}(n;\mathbb{F})$. Let $q = rmn$ and $\varphi : \mathbb{R}^q \longrightarrow \operatorname{Mat}(m,n;\mathbb{F})$ denote the projection with respect to the canonical basis, i.e.
$x = \left(x_{11}^{(1)},\ldots,x_{11}^{(r)},\ldots,x_{1n}^{(r)},\ldots,x_{mn}^{(r)} \right)' \in \mathbb{R}^q$ is mapped onto $X = (x_{kl})$, where $x_{kl} = \sum_{j=1}^{r} x_{kl}^{(j)} e_j$. Next choose $F \in \operatorname{Mat}(q;\mathbb{R})$ such that $\varphi(F\mathbb{Z}^q) = \Lambda$, hence $|\det F| = \operatorname{vol}\Lambda$. Put

$$\psi : \mathbb{R}^q \longrightarrow \mathbb{C}\ ,$$
$$x \longmapsto e^{-\pi\tau(S[\varphi(Fx)+P],Y)\ +\ 2\pi i\tau(Q,\varphi(Fx)+P)}\ .$$

Since φ is linear, we obtain the absolute and locally uniform convergence of the series

$$\sum_{g\in\mathbb{Z}^q} \psi(g+x)\ ,\quad x \in \mathbb{R}^q\ ,$$

by Theorem 1.2. Now we apply the POISSON summation formula (cf.[58], VII.2.6) to achieve

$$\Theta_{P,Q}(iY,S;\Lambda) = \sum_{g\in\mathbb{Z}^q} \psi(g)$$
$$= \sum_{h\in\mathbb{Z}^q} \int_{\mathbb{R}^q} \psi(x)\ e^{-2\pi i h'x}\ dx$$

Substituting $y = Fx$ yields

$$|\det F|\ \Theta_{P,Q}(iY,S;\Lambda)$$
$$= \sum_{h\in\mathbb{Z}^q} \int_{\mathbb{R}^q} e^{-\pi\tau(S[\varphi(y)+P],Y)\ +\ 2\pi i\tau(Q,\varphi(y)+P)\ -\ 2\pi i(F'^{-1}h)'y}\ dy\ .$$

One has $(F'^{-1}h)'y = \tau(\varphi(F'^{-1}h),\varphi(y))$. If h runs through $\mathbb{Z}^q$, then $\varphi(F'^{-1}h)$ runs through Λ^τ because of the definition of F . Thus we obtain

$$(\mathrm{vol}\ \Lambda)\ \Theta_{P,Q}(iY,S;\Lambda)$$
$$= \sum_{H\in\Lambda^\tau} \int_{\mathrm{Mat}(m,n;\mathbb{F})} e^{-\pi\tau(S[U+P],Y) + 2\pi i\tau(Q-H,U) + 2\pi i\tau(Q,P)}\, dU .$$

Proposition 1.1 yields

$$-\tau(S[U+P],Y) + 2i\tau(Q-H,U) + 2i\tau(Q,P)$$
$$= -\tau(S[U+P+iS^{-1}(H-Q)Y^{-1}],Y) + 2i\tau(H,P) - \tau(S^{-1}[H-Q],Y^{-1}) .$$

Finally we put $W := P + iS^{-1}(H-Q)Y^{-1} \in \mathrm{Mat}(m,n;\mathbb{F})_{\mathbb{C}}$ and apply Lemma 2.1 to achieve

$$(\mathrm{vol}\ \Lambda)\ \Theta_{P,Q}(iY,S;\Lambda)$$
$$= \sum_{H\in\Lambda^\tau} e^{-\pi\tau(S^{-1}[H-Q],Y^{-1}) + 2\pi i\tau(H,P)} \int_{\mathrm{Mat}(m,n;\mathbb{F})} e^{-\pi\tau(S[U+W],Y)}\, dU$$
$$= (\det S)^{-\frac{1}{2}rn}\ (\det Y)^{-\frac{1}{2}rm} \sum_{H\in\Lambda^\tau} e^{-\pi\tau(S^{-1}[H-Q],Y^{-1}) + 2\pi i\tau(H,P)}$$
$$= (\det S)^{-\frac{1}{2}rn}\ (\det Y)^{-\frac{1}{2}rm}\ e^{2\pi i\tau(Q,P)}\ \Theta_{-Q,P}(iY^{-1},S^{-1};\Lambda^\tau) .$$

□

In the special case $\Lambda = \Lambda(\mathbb{F})$ the preceding theorem and Corollary 1.7 lead to

<u>Corollary 2.3.</u> a) <u>Let</u> $\mathbb{F} = \mathbb{R},\mathbb{C}$, <u>then</u>

$$(\det \tfrac{1}{i}Z)^{\frac{1}{2}rm}\ (\det S)^{\frac{1}{2}rn}\ \Theta_{P,Q}(Z,S;\Lambda(\mathbb{F}))$$
$$= e^{2\pi i\tau(P,Q)}\ \Theta_{-Q,P}(-Z^{-1},S^{-1};\Lambda(\mathbb{F})) .$$

b) <u>If</u> $\alpha = e_1 + e_2$ <u>one has</u>

$$(\det \tfrac{1}{i}Z)^{2m}\ (\det S)^{2n}\ \Theta_{P,Q}(Z,S;\Lambda(\mathbb{H}))$$
$$= 2^{mn}\ e^{2\pi i\tau(P,Q)}\ \Theta_{-\alpha^{-1}Q,\bar{\alpha}P}(-Z^{-1},S^{-1}[\alpha I];\Lambda(\mathbb{H})) .$$

Definition. A matrix $S \in Pos(m;\mathbb{F})$ is called stable or, more precisely, $\mathbb{F}$-stable if S is $\mathbb{F}$-even and satisfies $\det S = (\mathrm{vol}\ \mathcal{O})^{-2m/r}$.

This class of matrices is called stable in order to indicate the connection with stable modular forms (cf. [21], p.279).

Consider U,V determined in Corollary 1.9.

Proposition 2.4. a) If S is $\mathbb{H}$-stable then $\check{S}[V]$ becomes $\mathbb{C}$-stable and $\hat{S}[U]$ becomes $\mathbb{R}$-stable.

b) Given a $\mathbb{C}$-stable matrix S then $\hat{S}$ turns out to be $\mathbb{R}$-stable.

Proof. The arising matrices are even in view of Proposition 1.12. Calculation of determinants completes the proof. □

Thus we obtain a necessary condition on the existence of stable matrices.

Corollary 2.5. If there exists an $\mathbb{F}$-stable $S \in Pos(m;\mathbb{F})$ one has $rm \equiv 0 \bmod 8$ as a necessary condition.

Proof. The assertion is well-known for $\mathbb{F} = \mathbb{R}$, cf [52], p.109, Th.8. The other cases follow by Proposition 2.4. □

Considering the so-called Theta-Nullwerte we use the abbreviation

$$\Theta(Z,S;\mathbb{F}) := \Theta^{(n)}(Z,S;\mathbb{F}) := \Theta_{0,0}(Z,S;\Lambda(\mathbb{F})) .$$

Theorem 2.6. Let $S \in Pos(m;\mathbb{F})$ be stable. Then $\Theta(\cdot,S;\mathbb{F})$ is a modular form on $H(n;\mathbb{F})$ of weight $\frac{1}{2}rm$ with FOURIER-expansion

$$\Theta(Z,S;\mathbb{F}) = \sum_{T \in Sym^{\tau}(n;\mathcal{O}),\, T \geq 0} \#(S,2T;\mathbb{F})\, e^{2\pi i \tau(T,Z)} , \quad Z \in H(n;\mathbb{F}).$$

One has

$$\Theta^{(n)}(\cdot,S;\mathbb{F})|\phi = \Theta^{(n-1)}(\cdot,S;\mathbb{F}) ,$$

where $\Theta^{(0)}(\cdot,S;\mathbb{F}) := 1$.

Proof. The FOURIER-expansion follows from the definition by a rearrangement, since S represents only even matrices in view of Proposition 1.10. To describe the effect of SIEGEL's ϕ-operator in suffices to consider $n = 1$ because of Lemma 1.14. According to Theorem 1.2 the limit may be distributed through the infinite series. Now

$$\lim_{y\to\infty} e^{-\pi y S[\mathfrak{g}]} = \begin{cases} 1, & \text{if } \mathfrak{g} = 0 \\ 0, & \text{if } \mathfrak{g} \neq 0 \end{cases}$$

leads to $\Theta^{(1)}(\cdot,S;\mathbb{F})|\phi = \lim_{y\to\infty} \Theta(iy,S;\mathbb{F}) = 1$.

Corollary 1.7 yields $\Theta(Z[U],S;\mathbb{F}) = \Theta(Z,S;\mathbb{F})$ for all $U \in GL(n;\mathcal{O})$. We obtain $\Theta(Z+H,S;\mathbb{F}) = \Theta(Z,S;\mathbb{F})$ for all $H \in Sym(n;\mathcal{O})$ from Proposition 1.11. If $\mathbb{F} = \mathbb{R},\mathbb{C}$, one has $S \in GL(m;\mathcal{O})$. Because of $\frac{1}{2}rm \equiv 0 \bmod 4$ according to the corollary above Corollary 2.3 and Corollary 1.7 yield

$$\Theta(-Z^{-1},S;\mathbb{F}) = (\det Z)^{\frac{1}{2}rm}\,\Theta(Z,S;\mathbb{F})$$

for $\mathbb{F} = \mathbb{R},\mathbb{C}$. Defining $\alpha := e_1 + e_2$ we obtain

$$\Theta(-Z^{-1},S^{-1}[\alpha I];\mathbb{H}) = (\det Z)^{2m}\,\Theta(Z,S;\mathbb{H})$$

from Corollary 2.3, since m is even. One has $W := \alpha^{-1}S \in Mat(m;\mathcal{O})$ in view of Proposition 1.10 and Lemma I.1.15. But $\det S = 2^{\frac{1}{2}m}$ implies $W \in GL(m;\mathcal{O})$. Because of $S^{-1}[\alpha I][W] = S$ Corollary 1.7 leads to

$$\Theta(-Z^{-1},S;\mathbb{H}) = (\det Z)^{2m}\,\Theta(Z,S;\mathbb{H}) .$$

Lemma III.1.7 completes the proof. □

Now we are going to construct $\mathbb{F}$-stable matrices. It immediately follows from the definition that

$$S_{\mathbb{H}} := \begin{pmatrix} 2 & e_1+e_2 \\ e_1-e_2 & 2 \end{pmatrix}$$

is $\mathbb{H}$-stable. Proposition 2.4 leads to the $\mathbb{C}$-stable matrix

$$S_{\mathbb{C}} := \begin{pmatrix} 2 & e_1-e_2 & e_1 & -e_2 \\ e_1+e_2 & 2 & e_1 & e_1-e_2 \\ e_1 & e_1 & 2 & e_1-e_2 \\ e_2 & e_1+e_2 & e_1+e_2 & 2 \end{pmatrix}$$

and to the $\mathbb{R}$-stable matrix

$$S_{\mathbb{R}} := \begin{pmatrix} 2 & 1 & 1 & 1 & 1 & 0 & 1 & 1 \\ 1 & 2 & 0 & 0 & 1 & 1 & 1 & 0 \\ 1 & 0 & 2 & 0 & 0 & -1 & 1 & 0 \\ 1 & 0 & 0 & 2 & 0 & 0 & 0 & 1 \\ 1 & 1 & 0 & 0 & 2 & 1 & 1 & 1 \\ 0 & 1 & -1 & 0 & 1 & 2 & 0 & 0 \\ 1 & 1 & 1 & 0 & 1 & 0 & 2 & 0 \\ 1 & 0 & 0 & 1 & 1 & 0 & 0 & 2 \end{pmatrix}.$$

Thus we have also shown that the condition stated in Corollary 2.5 is sufficient, too, i.e. given $m > 0$ satisfying $rm \equiv 0 \bmod 8$ there exists an $\mathbb{F}$-stable $S \in \mathrm{Pos}(m;\mathbb{F})$.

Especially, Theorem 2.6 and Corollary 1.9 lead to

<u>Corollary 2.7.</u> $\Theta(\cdot,S_{\mathbb{F}};\mathbb{F})$ <u>is a modular form on</u> $H(n;\mathbb{F})$ <u>of weight</u> 4 . <u>One has</u>

$$\Theta(\cdot,S_{\mathbb{H}};\mathbb{H})\big|_{H(n;\mathbb{C})} = \Theta(\cdot,S_{\mathbb{C}};\mathbb{C}) ,$$

$$\Theta(\cdot,S_{\mathbb{C}};\mathbb{C})\big|_{H(n;\mathbb{R})} = \Theta(\cdot,S_{\mathbb{R}};\mathbb{R}) .$$

Given $k \equiv 0 \bmod 4$ let $[\Gamma(n;\mathbb{F}),k]_{\Theta}$ denote the subspace of $[\Gamma(n;\mathbb{F}),k]$ generated by the theta-series $\Theta(\cdot,S;\mathbb{F})$, where $S \in \mathrm{Pos}(m;\mathbb{F})$, $m = \frac{2k}{r}$, is $\mathbb{F}$-stable.

Corollary 1.9 and Proposition 2.4 yield

<u>Corollary 2.8.</u> <u>Given</u> $f \in [\Gamma(n;\mathbb{H}),k]_{\Theta}$ <u>then</u> $f\big|_{H(n;\mathbb{C})} \in [\Gamma(n;\mathbb{C}),k]_{\Theta}$. <u>If</u> $f \in [\Gamma(n;\mathbb{C}),k]_{\Theta}$ <u>then</u> $f\big|_{H(n;\mathbb{R})}$ <u>belongs to</u> $[\Gamma(n;\mathbb{R}),k]_{\Theta}$.

If S is $\mathbb{F}$-stable, then $S[U]$, $U \in GL(m;\mathcal{O})$, turns out to be $\mathbb{F}$-stable, too. Let $st(m;\mathbb{F})$ denote the number of equivalence classes of stable matrices $S \in \mathrm{Pos}(m;\mathbb{F})$. One has

$$st(m;\mathbb{F}) < c_m(N;\mathbb{F}) < \infty \quad , \; N = (\mathrm{vol}\ \mathcal{O})^{-2m/r} ,$$

in view of Theorem I.5.2.

Lemma 2.9. *If* $k \equiv 0 \bmod 4$ *and* $m = \frac{2k}{r}$, *one has*

$$\dim\,[\Gamma(n;\mathbb{F}),k]_\Theta \leq st(m;\mathbb{F}) .$$

On condition that $n \geq m$, *even*

$$\dim\,[\Gamma(n;\mathbb{F}),k]_\Theta = st(m;\mathbb{F})$$

holds.

Proof. If two stable matrices are equivalent the corresponding theta-series coincide in view of Corollary 1.7. Considering SIEGEL's ϕ-operator and Theorem 2.6 it suffices to demonstrate

$$\dim\,[\Gamma(m;\mathbb{F}),k]_\Theta = st(m;\mathbb{F}) .$$

Let $S_1,\ldots,S_t$, $t = st(m;\mathbb{F})$, denote representatives of the classes of stable $S \in Pos(m;\mathbb{F})$. Theorem I.5.4 yields

$$\#(S_j,S_l;\mathbb{F}) = \delta_{jl}\ \#(S_j,S_j;\mathbb{F}) , \quad 1 \leq j,l \leq t .$$

We obtain the linear independence of $\Theta(\cdot,S_1;\mathbb{F}),\ldots,\Theta(\cdot,S_t;\mathbb{F})$ from the FOURIER-expansion in Theorem 2.6. □

The class numbers of stable matrices can become very large. SERRE [52],p.55, states

$$st(8;\mathbb{R}) = 1 , \quad st(16;\mathbb{R}) = 2 , \quad st(24;\mathbb{R}) = 24 \quad \text{and}$$
$$st(32;\mathbb{R}) > 8\cdot 10^7 .$$

QUEBBEMANN [47] shows that

$$st(2;\mathbb{H}) = 1 , \quad st(4;\mathbb{H}) = 1 , \quad st(6;\mathbb{H}) = 2 .$$

But the result "$st(2;\mathbb{H}) = 1$" can also easily be derived from the explicit description of the reduced 2×2 matrices (cf. remarks below Theorem I.4.12), since one has $c_2(2;\mathbb{H}) = 2$.

§3 Theta-series with respect to even quadratic forms

In this chapter we examine the behavior of theta-series with respect to even quadratic forms under modular transformations. Especially the action on the characteristics will be described explicitly. The corresponding results on SIEGEL's theta-series are due to ANDRIANOV and MALOLETKIN [2]. But we won't use the KRONECKER-product, instead of this we proceed as KRAZER [39] did using Theorem 1.4 and the theta-transformation-formula.

First we study the action of the modular group on the characteristics.

Lemma 3.1. Let $S \in Pos(m;\mathbb{F})$ be fixed. Given $M = \begin{pmatrix} A & B \\ C & D \end{pmatrix} \in Sp(n;\mathbb{F})$ and $(P,Q) \in Mat(m,n;\mathbb{F})^2_{\mathbb{C}}$ we define

$$M * (P,Q) = (P\bar{D}'-S^{-1}Q\bar{C}', Q\bar{A}'-SP\bar{B}') ,$$

$$\kappa(M;(P,Q)) = e^{\pi i\tau(S[P],\bar{D}'B) + \pi i\tau(S^{-1}[Q],\bar{A}'C) - 2\pi i\tau(\bar{P}'Q,\bar{B}'C)} .$$

Suppose that $M_1, M_2 \in Sp(n;\mathbb{F})$ then the following holds:

a) $$(M_1M_2) * (P,Q) = M_1 * (M_2 * (P,Q)) ,$$

b) $$\kappa(M_1M_2;(P,Q)) = \kappa(M_1;M_2*(P,Q))\ \kappa(M_2;(P,Q)) .$$

Proof. a) Given M_1, M_2 in standard form one has

$$\begin{aligned} M_1 * (M_2 * (P,Q)) &= M_1 * (P\bar{D}_2'-S^{-1}Q\bar{C}_2', Q\bar{A}_2'-SP\bar{B}_2') \\ &= \Big((P\bar{D}_2'-S^{-1}Q\bar{C}_2')\bar{D}_1' - S^{-1}(Q\bar{A}_2'-SP\bar{B}_2')\bar{C}_1' , \\ &\qquad (Q\bar{A}_2'-SP\bar{B}_2')\bar{A}_1' - S(P\bar{D}_2'-S^{-1}Q\bar{C}_2')\bar{B}_1'\Big) \\ &= \Big(P\,\overline{(D_1D_2+C_1B_2)}' - S^{-1}Q\,\overline{(D_1C_2+C_1A_2)}' , \\ &\qquad Q\,\overline{(A_1A_2+B_1C_2)}' - SP\,\overline{(A_1B_2+B_1D_2)}'\Big) \\ &= (M_1M_2) * (P,Q) . \end{aligned}$$

b) We compute that $\kappa(M_1M_2;(P,Q)) = e^{\pi i\alpha}$, where

$$\begin{aligned} \alpha = {} & \tau(S[P], \overline{(D_1D_2+C_1B_2)}'(A_1B_2+B_1D_2)) \\ & + \tau(S^{-1}[Q], \overline{(A_1A_2+B_1C_2)}'(C_1A_2+D_1C_2)) \end{aligned}$$

$$- 2\tau(\bar{P}'Q, \overline{(A_1B_2+B_1D_2)}'(C_1A_2+D_1C_2)) \;.$$

On the other hand one has $\kappa(M_1;M_2*(P,Q))\ \kappa(M_2;(P,Q)) = e^{\pi i\beta}$, where

$$\beta = \tau(S[P\bar{D}_2'-S^{-1}Q\bar{C}_2'],\bar{D}_1'B_1) + \tau(S^{-1}[Q\bar{A}_2'-SP\bar{B}_2'],\bar{A}_1'C_1)$$
$$- 2\tau(\overline{(P\bar{D}_2'-S^{-1}Q\bar{C}_2')}'(Q\bar{A}_2'-SP\bar{B}_2'),\bar{B}_1'C_1)$$
$$+ \tau(S[P],\bar{D}_2'B_2) + \tau(S^{-1}[Q],\bar{A}_2'C_2) - 2\tau(\bar{P}'Q,\bar{B}_2'C_2) \;.$$

Proposition 1.1 yields

$$\beta = \tau(S[P],W_1) + \tau(S^{-1}[Q],W_2) - 2\tau(\bar{P}'Q,W_3) \;,$$

where

$$W_1 = \bar{D}_2'\bar{D}_1'B_1D_2 + \bar{B}_2'\bar{A}_1'C_1B_2 + \bar{B}_2'\bar{C}_1'B_1D_2 + \bar{D}_2'\bar{B}_1'C_1B_2 + \bar{D}_2'B_2 \;,$$
$$W_2 = \bar{C}_2'\bar{D}_1'B_1C_2 + \bar{A}_2'\bar{A}_1'C_1A_2 + \bar{A}_2'\bar{C}_1'B_1C_2 + \bar{C}_2'\bar{B}_1'C_1A_2 + \bar{A}_2'C_2 \;,$$
$$W_3 = \bar{D}_2'\bar{B}_1'D_1C_2 + \bar{B}_2'\bar{A}_1'C_1A_2 + \bar{D}_2'\bar{B}_1'C_1A_2 + \bar{B}_2'\bar{C}_1'B_1C_2 + \bar{B}_2'C_2 \;.$$

Hence we obtain $\alpha = \beta$ from Lemma II.1.1.

□

First we consider the special case that S is $\mathbb{F}$-stable.

<u>Theorem 3.2.</u> <u>Let</u> $S \in Pos(m;\mathbb{F})$ <u>be</u> $\mathbb{F}$-<u>stable</u>. <u>Given</u> $M \in \Gamma(n;\mathbb{F})$, $Z \in H(n;\mathbb{F})$ <u>and</u> $P,Q \in Mat(m,n;\mathbb{F})_{\mathbb{C}}$ <u>one has</u>

$$\Theta_{P,Q}(Z,S;\Lambda(\mathbb{F})) = (\det M\{Z\})^{-\frac{1}{2}rm}\ \kappa(M;(P,Q))\ \Theta_{M*(P,Q)}(M\langle Z\rangle,S;\Lambda(\mathbb{F})).$$

<u>Proof.</u> Because of Lemma 3.1 and Theorem II.1.7 it suffices to prove the assertion for all matrices M belonging to the set of generators of $\Gamma(n;\mathbb{F})$ quoted in Theorem II.2.3. If $M = [\bar{U}',U^{-1}]$, $U \in GL(n;\mathcal{O})$, consider Corollary 1.7 and notice that $\frac{1}{2}rm \equiv 0 \bmod 4$ in view of Corollary 2.5. If $M = \begin{pmatrix} I & T \\ 0 & I \end{pmatrix}$, $T \in Sym(n;\mathcal{O})$, Proposition 1.11 yields the assertion. In the case $M = J$ use Corollary 2.3 and Corollary 1.7.

□

Next we state some auxiliary assertions, but first remember that we call matrices $P,Q \in Mat(m,n;\mathbb{F})$ <u>congruent</u> <u>modulo</u> $q\mathcal{O}$ ($q \in \mathbb{N}$) and write $P \equiv Q \bmod q\mathcal{O}$ if $P-Q \in Mat(m,n;q\mathcal{O})$ holds.

If Λ is a lattice in $Mat(m,n;\mathbb{F})$ then $P \in \Lambda \bmod q$ means that P runs through a set of representatives of the cosets $\Lambda/q\Lambda$.

Let $\Lambda(\mathbb{F}) := \mathrm{Mat}(m,n;\mathcal{O}(\mathbb{F}))$ again denote the lattice of integral $m \times n$ matrices. Considering KRAZER [31],II,§1, we prove

Proposition 3.3. *Suppose that* $W \in \mathrm{Mat}(n;\mathcal{O})$ *has maximal rank and* $t \in \mathbb{N}$ *satisfying* $tW^{-1} \in \mathrm{Mat}(n;\mathcal{O})$. *Let* $\lambda = \lambda(W,t)$ *denote the number of solutions* $\mathfrak{g} \in \mathcal{O}^n \bmod t$ *of the congruence* $tW^{-1}\mathfrak{g} \equiv 0 \bmod t\mathcal{O}$.

a) *Given a function* $\varphi : \mathrm{Mat}(m,n;\mathbb{F}) \longrightarrow \mathbb{C}$ *then*

$$\lambda^m \sum_{G \in \Lambda(\mathbb{F})} \varphi(G) = \sum_{\substack{H \in \Lambda(\mathbb{F}) \bmod t \\ G \in \Lambda(\mathbb{F})}} \varphi(GW+H)$$

holds, if one of the two series converges absolutely.

b) $$\lambda(W,t) = t^{rn} \, |\det \hat{W}|^{-1} .$$

Proof. a) If one series converges absolutely, so does the other one. The assertion does not depend on the choice of the representatives of the cosets. Let G run through $\Lambda(\mathbb{F})$ and $H \in \Lambda(\mathbb{F}) \bmod t$. Then $K := GW+H$ runs through $\Lambda(\mathbb{F})$ exactly λ^m-times by definition of λ . Because of the absolute convergence we may rearrange the series.

b) $\lambda(W,t)$ does not change, if we multiply W by unimodular matrices. In view of Theorem I.2.3 we may suppose that W is diagonal. Hence it suffices to consider $n = 1$, i.e. $w \in \mathcal{O} - \{0\}$.

If $\mathbb{F} = \mathbb{R}$ the assertion is obvious. Considering $\mathbb{F} = \mathbb{C}$ we observe that $\lambda(w,t) = \lambda(\hat{w},t)$ and obtain the value of λ from the case $\mathbb{F} = \mathbb{R}$. Let $\mathbb{F} = \mathbb{H}$ and u as defined in §1. Notice that $u^{-1}\hat{w}u \in \mathrm{Mat}(4;\mathbb{Z})$ and $\lambda(w,t) = \lambda(u^{-1}\hat{w}u,t)$. Hence the case $\mathbb{F} = \mathbb{R}$ completes the proof. □

Definition. Let $S \in \mathrm{Pos}(m;\mathbb{F})$ be $\mathbb{F}$-even. We say that S is of *level* q $(q \in \mathbb{N})$ if $q(\mathrm{vol}\,\mathcal{O}\cdot S)^{-1}$ becomes $\mathbb{F}$-even.

We state some elementary properties.

Proposition 3.4. *Let* $S \in \mathrm{Pos}(m;\mathbb{F})$ *be* $\mathbb{F}$*-even.*

a) S *is of level* 1 *if and only if* S *is* $\mathbb{F}$*-stable.*

b) *If* $\mathbb{F} = \mathbb{H}$ *and* $\alpha = e_1 + e_2$ *the following assertions are equivalent:*

(i) S *is of level* q.

(ii) $qS^{-1}[\alpha I]$ *is* $\mathbb{F}$*-even*.

(iii) $qS^{-1}(\alpha I) \in \mathrm{Mat}(m;\mathcal{O})$.

Proof. a) The cases $\mathbb{F} = \mathbb{R}, \mathbb{C}$ are trivial. If $\mathbb{F} = \mathbb{H}$, we have $\alpha^{-1}S \in \mathrm{Mat}(m;\mathcal{O})$, hence $\det S \geq 2^{\frac{1}{2}m}$. In view of $\mathrm{vol}\ \mathcal{O} = \frac{1}{2}$ the assertion is obvious.

b) Since $e_1 + e_2$ is invariant according to Lemma I.1.5, (i) and (ii) are equivalent. (ii) and (iii) are equivalent in view of the description of even matrices in Proposition 1.10 and Lemma I.1.15. □

Given $q \in \mathbb{N}$ let $\Gamma(n;\mathbb{F})\langle q\rangle$ denote the theta-group, i.e.

$$\Gamma(n;\mathbb{F})\langle q\rangle = \left\{ \begin{pmatrix} A & B \\ C & D \end{pmatrix} \in \Gamma(n;\mathbb{F}) \ ;\ C \equiv 0 \bmod q\mathcal{O} \right\}.$$

We apply Theorem 1.4 to derive

Proposition 3.5. Suppose that $S \in \mathrm{Pos}(m;\mathbb{F})$ *is even and of level* q, $Z \in H(n;\mathbb{F})$, $M = \begin{pmatrix} A & B \\ C & D \end{pmatrix} \in \Gamma(n;\mathbb{F})\langle q\rangle$ *and* $P, Q \in \mathrm{Mat}(m,n;\mathbb{F})_{\mathbb{C}}$. *Then there exists* $\chi \in \mathbb{C}$ *such that*

$$e^{\pi i \tau(S[W],(AZ+B)\bar{A}')}\ \Theta_{P,Q+SW(AZ+B)}(Z,S;\Lambda(\mathbb{F}))$$

$$= \chi\, \Theta_{P_0,Q_0+W}(-M\langle Z\rangle^{-1}, S^{-1}; \Lambda(\mathbb{F})^{\tau})$$

holds for all $W \in \mathrm{Mat}(m,n;\mathbb{F})_{\mathbb{C}}$, *where*

$$P_0 = SP\bar{B}' - Q\bar{A}' \ ,\ Q_0 = P\bar{D}' - S^{-1}Q\bar{C}'.$$

Proof. Put $\varphi(W) := e^{\pi i \tau(S[W],(AZ+B)\bar{A}')}$, $\psi(W) := \Theta_{P,Q+SW(AZ+B)}(Z,S;\Lambda(\mathbb{F}))$. If $H \in \Lambda(\mathbb{F})$, Proposition 1.1 yields

$$\varphi(W+H) = \varphi(W)\ e^{\pi i \tau(S[H],AZ\bar{A}') + 2\pi i \tau(\bar{H}'SW,(AZ+B)\bar{A}')}$$

because of $\tau(S[H],B\bar{A}') \equiv 0 \bmod 2$, since $B\bar{A}' \in \mathrm{Sym}(n;\mathcal{O})$ and S is even. One has

$$\psi(W+H) = \Theta_{P,Q_1}(Z,S;\Lambda(\mathbb{F})),$$

where $Q_1 = Q + SW(AZ+B) + SHAZ + SHB$. Put $Q_2 = Q + SW(AZ+B) + SHAZ$.

Because of $SHB \in \Lambda(\mathbb{F})^{\tau}$ and $HA \in \Lambda(\mathbb{F})$ Proposition 1.3 leads to

$$\Theta_{P,Q_1}(Z,S;\Lambda(\mathbb{F})) = e^{2\pi i\tau(P,SHB)}\ \Theta_{P,Q_2}(Z,S;\Lambda(\mathbb{F}))$$

$$= \psi(W)\ e^{2\pi i\tau(P,SHB) - \pi i\tau(S[HA],Z) - 2\pi i\tau(Q+SW(AZ+B),HA)} .$$

In view of $(AZ+B)\bar{A}' \in \mathrm{Sym}(n;\mathbb{F}) \otimes_{\mathbb{R}} \mathbb{C}$ we obtain

$$\varphi(W+H)\ \psi(W+H) = e^{2\pi i\tau(H,SP\bar{B}'-Q\bar{A}')}\ \varphi(W)\ \psi(W)$$

from Proposition 1.1.

Given $K \in \Lambda(\mathbb{F})^{\tau}$ one has

$$\varphi(W-S^{-1}KM\langle Z\rangle^{-1})$$

$$= \varphi(W)\ e^{\pi i\tau(S^{-1}[K],(CZ+D)\bar{A}'M\langle Z\rangle^{-1}) - 2\pi i\tau(\bar{K}'W,(CZ+D)\bar{A}')} .$$

Putting $Q_3 = Q + SW(AZ+B) - KCZ - KD$ yields

$$\psi(W-S^{-1}KM\langle Z\rangle^{-1}) = \Theta_{P,Q_3}(Z,S;\Lambda(\mathbb{F})) .$$

Now $KD \in \Lambda(\mathbb{F})^{\tau}$ leads to

$$\Theta_{P,Q_3}(Z,S;\Lambda(\mathbb{F})) = e^{-2\pi i(P,KD)}\ \Theta_{P,Q_4}(Z,S;\Lambda(\mathbb{F}))$$

in view of Proposition 1.3, where $Q_4 = Q + SW(AZ+B) - KCZ$. Next $S^{-1}KC \in \Lambda(\mathbb{F})$ follows from $C \equiv 0 \bmod q\mathcal{O}$. Apply Proposition 1.3 to achieve

$$\Theta_{P,Q_3}(Z,S;\Lambda(\mathbb{F}))$$

$$= \psi(W)\ e^{-\pi i\tau(S^{-1}[K],(CZ+D)\bar{C}')-2\pi i\tau(K,P\bar{D}'-S^{-1}Q\bar{C}')+2\pi i\tau(\bar{K}'W,C\overline{(AZ+B)}')} ,$$

since $\tau(S^{-1}[K],D\bar{C}') \equiv 0 \bmod 2$. Lemma II.1.1 yields

$$\tau(\bar{K}'W,(CZ+D)\bar{A}' - C\overline{(AZ+B)}') = \tau(K,W) ,$$

$$(CZ+D)\bar{C}' - (CZ+D)\bar{A}'\,M\langle Z\rangle^{-1} = -M\langle Z\rangle^{-1} .$$

Summarizing we have

$$\varphi(W-S^{-1}K\,M\langle Z\rangle^{-1})\ \ \psi(W-S^{-1}K\,M\langle Z\rangle^{-1})$$

$$= \varphi(W)\ \psi(W)\,e^{-\pi i\tau(S^{-1}[K],-M\langle Z\rangle^{-1}) - 2\pi i\tau(P\bar{D}'-S^{-1}Q\bar{C}'+W,K)} .$$

As φ and ψ are holomorphic, Theorem 1.4 completes the proof. □

We are able to compute χ explicitly if A is invertible. But the invertibility of A is not a condition at all: If S of level 1 consider Theorem 3.2. If S is of level $q > 1$ then A has maximal rank, since M is unimodular.

Theorem 3.6. *Suppose that* $S \in \mathrm{Pos}(m;\mathbb{F})$ *is even and of level* $q > 1$ *and* $rm \equiv 0 \bmod 2$. *Given* $M = \begin{pmatrix} A & B \\ C & D \end{pmatrix} \in \Gamma(n;\mathbb{F})\langle q\rangle$ *and* $t \in \mathbb{N}$ *such that* $tA^{-1} \in \mathrm{Mat}(n;\mathcal{O})$, *all* $Z \in H(n;\mathbb{F})$ *and* $P,Q \in \mathrm{Mat}(m,n;\mathbb{F})_{\mathbb{C}}$ *satisfy*

$$\Theta_{P,Q}(Z,S;\Lambda(\mathbb{F})) = \nu(M)(\det M\{Z\})^{-\frac{1}{2}rm}\,\kappa(M;(P,Q))\,\Theta_{M*(P,Q)}(M\langle Z\rangle,S;\Lambda(\mathbb{F})),$$

where $\nu(M)$ *is given by*

$$\nu(M) = t^{-rmn}(\det A)^{\frac{1}{2}rm}\sum_{K\in\Lambda(\mathbb{F}) \bmod t} e^{-\pi i\tau(S[K],A^{-1}B)}, \quad \textit{if } \mathbb{F}=\mathbb{R},\mathbb{C},$$

$$\nu(M) = t^{-4mn}(\det \hat{A})^{\frac{1}{2}m}\sum_{K\in\Lambda(\mathbb{H}) \bmod t} e^{-\pi i\tau(S[K],A^{-1}B)}, \quad \textit{if } \mathbb{F}=\mathbb{H}.$$

Proof. Proposition 3.5 yields

$$\chi\, e^{-2\pi i\tau(P_o,W)}\,\Theta_{P_o,Q_o+W}(-M\langle Z\rangle^{-1};S^{-1};\Lambda(\mathbb{F})^{\tau})$$
$$= e^{\pi i\tau(S[W],(AZ+B)A') - 2\pi i\tau(P_o,W)}\,\Theta_{P,Q+SW(AZ+B)}(Z,S;\Lambda(\mathbb{F})).$$

Let $C(m,n;\mathbb{F})$ denote a fundamental parallelotope of the lattice $\Lambda(\mathbb{F})$. Now integrate the equation above over $C(m,n;\mathbb{F})$. Then the left side coincides with

$$\chi\int_{C(m,n;\mathbb{F})}\sum_{H\in\Lambda(\mathbb{F})^{\tau}} e^{\pi i\tau(S^{-1}[H+P_o],-M\langle Z\rangle^{-1})+2\pi i\tau(Q_o,H+P_o)+2\pi i\tau(W,H)}\,dW.$$

Because of the locally uniform convergence in view of Theorem 1.2 we may distribute the integral through the infinite series. If $H \neq 0$ the arising integral equals 0. If $H = 0$ the integral equals

$$\chi(\mathrm{vol}\,\mathcal{O})^{mn}\,e^{\pi i\tau(S^{-1}[P_o],-M\langle Z\rangle^{-1}) + 2\pi i\tau(Q_o,P_o)}.$$

Thus we have

$$\chi(\mathrm{vol}\,\mathcal{O})^{mn}\,e^{2\pi i\tau(Q_o,P_o)}$$
$$= \int_{C(m,n;\mathbb{F})}\sum_{G\in\Lambda(\mathbb{F})} e^{\pi i\varphi(W,G)}\,dW,$$

where $\varphi(W,G) = \tau(S[G+P],Z) + 2\tau(Q+SW(AZ+B),G+P)$

$$+ \tau(S[W],(AZ+B)\bar{A}') - 2\tau(P_o,W) + \tau(S^{-1}[P_o],M<Z>^{-1}) .$$

A verification using Lemma II.1.1 and Proposition 1.1 yields

$$\varphi(W,G) = \tau(S[W+(G+P)A^{-1}-S^{-1}P_oR^{-1}],R) - \tau(S[G],A^{-1}B)$$
$$+ \tau(S[P],\bar{D}'B) + \tau(S^{-1}[Q],\bar{A}'C) - 2\tau(\bar{P}'Q,\bar{B}'C) ,$$

where $R = (AZ+B)\bar{A}' \in H(n;\mathbb{F})$. Since S is even, one computes that

$$\varphi(W,GA+K) \equiv \varphi(W+G,K) \bmod 2$$

holds for all $K \in \Lambda(\mathbb{F})$. Now apply Proposition 3.3 to obtain

$$\chi(\mathrm{vol}\ \mathcal{O})^{mn}\, e^{2\pi i(P_o,Q_o)}$$

$$= \int_{C(m,n;\mathbb{F})} \sum_{G\in\Lambda(\mathbb{F})} e^{\pi i\varphi(W,G)}\, dW$$

$$= \lambda^{-m} \int_{C(m,n;\mathbb{F})} \sum_{\substack{K\in\Lambda(\mathbb{F}) \bmod t\\ G\in\Lambda(\mathbb{F})}} e^{\pi i\varphi(W+G,K)}\, dW$$

$$= \lambda^{-m} \sum_{K\in\Lambda(\mathbb{F}) \bmod t} \int_{\mathrm{Mat}(m,n;\mathbb{F})} e^{\pi i\varphi(W,K)}\, dW .$$

Lemma 2.1 yields

$$\int_{\mathrm{Mat}(m,n;\mathbb{F})} e^{\pi i\tau(S[W+(K+P)A^{-1}-S^{-1}P_oR^{-1}],R)}\, dW$$

$$= (\det S)^{-\frac{1}{2}rn} (\det \tfrac{1}{i}R)^{-\frac{1}{2}rm} .$$

Hence it follows that

$$\chi = (\mathrm{vol}\ \mathcal{O})^{-mn} e^{-2\pi i\tau(P_o,Q_o)} \kappa(M;(P,Q)) (\det S)^{-\frac{1}{2}rn} (\det \tfrac{1}{i}R)^{-\frac{1}{2}rm}$$

$$\cdot\ \lambda^{-m} \sum_{K\in\Lambda(\mathbb{F}) \bmod t} e^{-\pi i\tau(S[K],A^{-1}B)} .$$

Next we apply Proposition 3.5 putting $W = 0$ and Theorem 2.2:

$$\Theta_{P,Q}(Z,S;\Lambda(\mathbb{F})) = \chi\, \Theta_{P_o,Q_o}(-M<Z>^{-1},S^{-1};\Lambda(\mathbb{F})^{\tau})$$

$$= \kappa(M;(P,Q))\, \Theta_{M*(P,Q)}(M<Z>,S;\Lambda(\mathbb{F}))\, (\det \tfrac{1}{i}M<Z>)^{\frac{1}{2}rm} (\det \tfrac{1}{i}R)^{-\frac{1}{2}rm}$$

$$\cdot\ \lambda^{-m} \sum_{K\in\Lambda(\mathbb{F}) \bmod t} e^{-\pi i\tau(S[K],A^{-1}B)} .$$

Proposition 3.3 yields $\lambda^{-m} = t^{-rmn} |\det \hat{A}|^{m}$. Hence the proof is complete if we notice that m is even for $\mathbb{F} = \mathbb{R}$. □

Now we examine the function ν arising above.

Theorem 3.7. *Suppose that $S \in \mathrm{Pos}(m;\mathbb{F})$ is even and of level $q > 1$ and $rm \equiv 0 \bmod 2$. Then ν is a character of the theta-group $\Gamma(n;\mathbb{F})\langle q\rangle$ satisfying $\nu^4 \equiv 1$. If $\mathbb{F} = \mathbb{R},\mathbb{H}$ or $\mathbb{F} = \mathbb{C}$ and $m \equiv 0 \bmod 2$ even $\nu^2 \equiv 1$ holds.*

Proof. Suppose that $M_o, M_1 \in \Gamma(n;\mathbb{F})\langle q\rangle$. Theorem 3.6 especially yields

$$\begin{aligned}
&\nu(M_oM_1)\,(\det M_oM_1\{Z\})^{-\frac{1}{2}rm}\;\Theta(M_oM_1\langle Z\rangle,S;\mathbb{F})\\
&= \Theta(Z,S;\mathbb{F})\\
&= \nu(M_1)(\det M_1\{Z\})^{-\frac{1}{2}rm}\;\Theta(M_1\langle Z\rangle,S;\mathbb{F})\\
&= \nu(M_o)\nu(M_1)\;(\det M_1\{Z\})^{-\frac{1}{2}rm}(\det M_o\{M_1\langle Z\rangle\})^{-\frac{1}{2}rm}\Theta(M_o\langle M_1\langle Z\rangle\rangle,S;\mathbb{F}).
\end{aligned}$$

Since $\Theta(\cdot,S;\mathbb{F})$ does not vanish identically, Theorem II.1.7 leads to

$$\nu(M_oM_1) = \nu(M_o)\;\nu(M_1)\;.$$

In order to demonstrate $\nu^4 \equiv 1$ it suffices to consider those M belonging to the set of generators of $\Gamma(n;\mathbb{F})\langle q\rangle$ quoted in Theorem II.4.2. Putting $t = 1$ the definition leads to

$$\nu\begin{pmatrix} I & T\\ O & I\end{pmatrix} = \nu\begin{pmatrix} I & O\\ qT & I\end{pmatrix} = 1 \qquad ,\ \text{whenever}\quad T \in \mathrm{Sym}(n;\mathcal{O})\ ,$$

$$\nu\begin{pmatrix} \bar{U}' & O\\ O & U^{-1}\end{pmatrix} = (\det U)^{-\frac{1}{2}rm} \qquad ,\ \text{whenever}\quad \mathbb{F} = \mathbb{R},\mathbb{C}\ ,$$

$$\nu\begin{pmatrix} \bar{U}' & O\\ O & U^{-1}\end{pmatrix} = 1 \qquad ,\ \text{whenever}\quad \mathbb{F} = \mathbb{H}\ ,\quad U \in GL(n;\mathcal{O}).$$

According to Lemma II.2.6 it suffices to consider $M = I \times K$, where $K = \begin{pmatrix} a & b\\ c & d\end{pmatrix} \in SL(2;\mathbb{Z})$, $c \equiv 0 \bmod q$. We may assume that $a > 0$ is odd, as we otherwise may consider $-K$ or $\begin{pmatrix} 1 & 1\\ o & 1\end{pmatrix}\begin{pmatrix} a & b\\ c & d\end{pmatrix} = \begin{pmatrix} a+c & b+d\\ c & d\end{pmatrix}$. Hence

$$\begin{aligned}
\nu(M) &= a^{-\frac{1}{2}rm(2n-1)} \sum_{G\in\Lambda(\mathbb{F}) \bmod a} e^{-\pi i\tau(S[G],A^{-1}B)}\\
&= a^{-\frac{1}{2}rm} \sum_{\mathfrak{g}\in\mathcal{O}^m \bmod a} e^{-\pi i a^{-1}bS[\mathfrak{g}]}\ .
\end{aligned}$$

Defining $T \in \mathrm{Pos}(rm;\mathbb{R})$ by $T := \hat{S}$, if $\mathbb{F} = \mathbb{R},\mathbb{C}$, and $T = \hat{S}[U]$, if $\mathbb{F} = \mathbb{H}$, where U was determined in Corollary 1.9, T is $\mathbb{R}$-even according to Proposition 1.12, hence

$$\nu(M) = a^{-\frac{1}{2}rm} \sum_{g \in \mathbb{Z}^{rm} \bmod a} e^{-\pi i a^{-1} b\, T[g]} .$$

Now $\nu(M) \in \{\pm 1\}$ holds according to EICHLER [15],§2*,3 . □

If $S \in \mathrm{Pos}(m;\mathbb{R})$, where m is odd, a corresponding result is due to CARLSSON and JOHANNSSEN [10].

<u>Theorem 3.8.</u> <u>Suppose that</u> $S \in \mathrm{Pos}(m;\mathbb{F})$ <u>is even and of level</u> q_1 <u>and</u> $rm \equiv 0 \bmod 2$. <u>Let</u> $P,Q \in \mathrm{Mat}(m,n;\mathbb{F})$ <u>be rational and</u> $q_2 \in \mathbb{N}$ <u>such that</u> $q_2 P \in \Lambda(\mathbb{F})$ <u>and</u> $q_2 Q \in \Lambda(\mathbb{F})^{\tau}$. <u>Then</u> $\Theta_{P,Q}(\cdot,S;\Lambda(\mathbb{F}))$ <u>is a modular form on</u> $H(n;\mathbb{F})$ <u>of weight</u> $\frac{1}{2}rm$ <u>with respect to the principal congruence subgroup of level</u> $q = q_1 q_2{}^2$ <u>and to the character</u> ν .

<u>Proof.</u> Given $M \in \Gamma(n;\mathbb{F})[q]$ one has $\kappa(M;(P,Q)) = 1$ by definition of κ and by the choice of q . If S is of level 1 use Theorem 3.2 and put $\nu \equiv 1$. In the general case Theorem 3.6 yields

$$\Theta_{P,Q}(Z,S;\Lambda(\mathbb{F})) = \nu(M) \ (\det M\{Z\})^{-\frac{1}{2}rm}\ \Theta_{M*(P,Q)}(M\langle Z\rangle,S;\Lambda(\mathbb{F})) .$$

Finally one computes that

$$\Theta_{M*(P,Q)}(M\langle Z\rangle,S;\Lambda(\mathbb{F})) = \Theta_{P,Q}(M\langle Z\rangle,S;\Lambda(\mathbb{F}))$$

by means of Proposition 1.3 and the definition of q .

□

In the special case $P = Q = 0$ one has:

Suppose that $S \in \mathrm{Pos}(m;\mathbb{F})$ is even and of level q and $rm \equiv 0 \bmod 2$, then

$$\Theta(\cdot,S;\mathbb{F}) \in [\Gamma(n;\mathbb{F})\langle q\rangle,\tfrac{1}{2}rm,\nu] .$$

§4 Singular modular forms

Singular modular forms are defined in contrast to cusp forms. We are going to prove the structure theorem saying that the space of singular modular forms coincides with the space of all modular forms and is generated by theta-series, whenever the weight is small in proportion to the degree. Singular modular forms were investigated by FREITAG [19],[20] and RESNIKOFF [48],[49]. The proofs are adopted from [21],IV,§5, A.IV .

Definition. A modular form $f \in [\Gamma(n;\mathbb{F}),k]$ possessing the FOURIER-expansion

$$f(Z) = \sum_{T\in Sym^{\tau}(n;\mathcal{O}),T\geq 0} \alpha(T)\, e^{2\pi i\tau(T,Z)} \quad , \; Z \in H(n;\mathbb{F}) \; ,$$

is called singular if $\alpha(T) = 0$ for every $T > 0$.

Singular modular forms are characterized by the condition "$\alpha(T) \neq 0 \Rightarrow \det T = 0$" whereas cusp forms can be defined by the property "$\alpha(T) \neq 0 \Rightarrow \det T \neq 0$" .

In the case $n = 1$ all singular modular forms are evidently constant. All modular forms of weight 0 - namely the constant maps - are singular.

But there also exist non-trivial examples. Let $S \in Pos(m;\mathbb{F})$ be $\mathbb{F}$-stable, then $\Theta^{(n)}(\cdot,S;\mathbb{F})$ belongs to $[\Gamma(n;\mathbb{F}),\frac{1}{2}rm]$ in view of Theorem 2.6. The numbers of representations $\#(S,2T;\mathbb{F})$, where $T \in Sym^{\tau}(n;\mathcal{O})$, $T \geq 0$, which arise in the FOURIER-expansion, can only be different from 0 if rank T does not exceed m . Hence $\Theta^{(n)}(\cdot,S;\mathbb{F})$ turns out to be singular if and only if $n > m$ holds. On the whole we now know all singular modular forms. But first we prove

Lemma 4.1. *Let $f \in [\Gamma(n;\mathbb{F}),k]$, $k > 0$, be a non-identically vanishing singular modular form with the FOURIER-expansion*

$$f(Z) = \sum_{T\in Sym^{\tau}(n;\mathcal{O}),T\geq 0} \alpha(T)\, e^{2\pi i\tau(T,Z)} \quad , \; Z \in H(n;\mathbb{F}) \; .$$

Then $k \equiv 0 \bmod 4$ *and* $m = \frac{2k}{r} < n$ *hold and there exists an* $\mathbb{F}$*-stable*

$S \in \mathrm{Pos}(m;\mathbb{F})$ <u>such that</u> $\alpha\begin{pmatrix}\frac{1}{2}S & 0\\ 0 & 0\end{pmatrix} \neq 0$.

<u>Proof.</u> If f were a cusp form it would follow that $f \equiv 0$. Hence we have $k \equiv 0 \bmod s(\mathbb{F})$ in view of Lemma III.2.13. Since f does not identically vanish, there exists a matrix $T \in \mathrm{Sym}^{\tau}(n;\mathcal{O})$ satisfying $T \geq 0$ and $\alpha(T) \neq 0$ such that rank $T = m$ becomes maximal under these conditions. As f is singular, we have $m < n$ and $k > 0$ implies that $m > 0$. In view of Proposition III.2.2 we can find $U \in GL(n;\mathcal{O})$ and a positive definite $S \in \mathrm{Sym}^{\tau}(m;\mathcal{O})$ such that $T[U] = \begin{pmatrix}S & 0\\ 0 & 0\end{pmatrix}$. Because of $\alpha(T) = \alpha\begin{pmatrix}S & 0\\ 0 & 0\end{pmatrix}$ we can choose S such that $\det S$ becomes minimal under the condition $\alpha\begin{pmatrix}S & 0\\ 0 & 0\end{pmatrix} \neq 0$. Let such an S be fixed.

The function $f([Z_1,Z_2])$, $Z_1 \in H(m;\mathbb{F})$, is developed into a FOURIER-series:

$$f([Z_1,Z_2]) = \sum_{T_1 \in \mathrm{Sym}^{\tau}(m;\mathcal{O}),\, T_1 \geq 0} \alpha_{T_1}(Z_2)\, e^{2\pi i \tau(T_1,Z_1)}, \quad Z_1 \in H(m;\mathbb{F}) .$$

Proposition III.1.10 yields that α_{T_1} is a modular form of degree $n-m$ and weight k given by

$$\alpha_{T_1}(Z_2) = \sum_{\substack{T_2 \in \mathrm{Sym}^{\tau}(n-m;\mathcal{O})\\ T_2 \geq 0}} \; \sum_{T = \begin{pmatrix}T_1 & T_{12}\\ * & T_2\end{pmatrix}} \alpha(T)\; e^{2\pi i \tau(T_2,Z_2)} ,$$

where the sum is only taken over those T_{12} for which $T = \begin{pmatrix}T_1 & T_{12}\\ * & T_2\end{pmatrix}$ in $\mathrm{Sym}^{\tau}(n;\mathcal{O})$ is positive semi-definite. We determine the special modular form $\alpha_S \in [\Gamma(n-m;\mathbb{F}),k]$. Let $T = \begin{pmatrix}S & T_{12}\\ * & T_2\end{pmatrix} \in \mathrm{Sym}^{\tau}(n;\mathcal{O})$ such that $\alpha(T) \neq 0$. Since rank S is maximal, there exist $U \in GL(n;\mathcal{O})$ and $T_0 \in \mathrm{Sym}^{\tau}(m;\mathcal{O})$, $T_0 \geq 0$, satisfying

$$\begin{pmatrix}S & T_{12}\\ * & T_2\end{pmatrix} = T = \begin{pmatrix}T_o & 0\\ 0 & 0\end{pmatrix}[U] ,$$

if we regard Proposition III.2.2. Let $U = \begin{pmatrix}G & *\\ * & *\end{pmatrix}$, where $G = G^{(m)}$. Especially, we have $S = T_0[G]$, hence $\det T_0 \neq 0$. From $\det S = \det T_o \det(\bar{G}'G)$ it follows that $\det T_o \leq \det S$. Since $\det S$ is minimal, we get $\det T_o = \det S$ and $G \in GL(m;\mathcal{O})$. Now

$\alpha\begin{pmatrix} S & 0 \\ 0 & 0 \end{pmatrix} = \alpha\begin{pmatrix} T_0 & 0 \\ 0 & 0 \end{pmatrix} = \alpha(T)$ leads to

$$\alpha_S(Z_2) = \alpha\begin{pmatrix} S & 0 \\ 0 & 0 \end{pmatrix} \sum_{\begin{pmatrix} S & T_{12} \\ * & T_2 \end{pmatrix} \sim \begin{pmatrix} S & 0 \\ 0 & 0 \end{pmatrix}} e^{2\pi i \tau(T_2, Z_2)} .$$

Let $U = \begin{pmatrix} U_1 & U_{12} \\ U_{21} & U_2 \end{pmatrix} \in GL(n;\mathcal{O})$ such that $U_1 = U_1^{(m)}$ and $\begin{pmatrix} S & 0 \\ 0 & 0 \end{pmatrix}[U]$ $= \begin{pmatrix} S & T_{12} \\ * & T_2 \end{pmatrix}$. It follows that $U_1 \in \mathrm{Aut}(S;\mathbb{F})$, $G := U_1^{-1}U_{12} \in \mathrm{Mat}(m,n-m;\mathcal{O})$ and $\begin{pmatrix} S & T_{12} \\ * & T_2 \end{pmatrix} = \begin{pmatrix} S & SG \\ * & S[G] \end{pmatrix}$. Since these matrices are mutually distinct, one calculates that

$$\alpha_S(Z_2) = \alpha\begin{pmatrix} S & 0 \\ 0 & 0 \end{pmatrix} \sum_{G \in \mathrm{Mat}(m,n-m;\mathcal{O})} e^{2\pi i \tau(S[G], Z_2)}$$

$$= \alpha\begin{pmatrix} S & 0 \\ 0 & 0 \end{pmatrix} \Theta(Z_2, 2S;\mathbb{F}) .$$

We have $\alpha\begin{pmatrix} S & 0 \\ 0 & 0 \end{pmatrix} \neq 0$. Thus $\Theta(\cdot, 2S;\mathbb{F})$ is a modular form of degree $n-m$ and weight k. By definition $2S$ is $\mathbb{F}$-even and Theorem 2.2 yields $\det S = (\mathrm{vol}\ \mathcal{O})^{-2m/r}$. Hence $2S$ is $\mathbb{F}$-stable and $k \equiv 0 \bmod 4$ holds according to Corollary 2.5.

□

The preceding lemma turns out to be the crucial tool in order to derive

<u>Theorem 4.2.</u> <u>Let</u> $f \in [\Gamma(n;\mathbb{F}),k]$, $k > 0$, <u>be a non-identically vanishing singular modular form. Then one has</u>

$$k \equiv 0 \bmod 4, k < \tfrac{1}{2}rn \quad \underline{\text{and}} \quad f \in [\Gamma(n;\mathbb{F}),k]_\Theta .$$

<u>Proof.</u> Regarding Lemma 4.1 it remains to prove that f is a linear combination of theta-series. Let $S_1, \ldots, S_t$ be a set of representatives of the equivalence classes of $\mathbb{F}$-stable matrices $S = S^{(m)}$. Put

$$g(Z) := f(Z) - \sum_{j=1}^{t} c_j\, \Theta(Z, S_j;\mathbb{F}) ,$$

where $c_j := \alpha\begin{pmatrix}\frac{1}{2}S_j & 0\\ 0 & 0\end{pmatrix} \#(S_j,S_j;\mathbb{F})^{-1}$. Because of $n > m$ together with f also g is singular. Let $\beta(T)$ denote the FOURIER-coefficients of g . One has $\beta\begin{pmatrix}\frac{1}{2}S & 0\\ 0 & 0\end{pmatrix} = 0$ for all $\mathbb{F}$-stable matrices $S = S^{(m)}$ according to the construction of g . Lemma 4.1 yields $g \equiv 0$, hence $f \in [\Gamma(n;\mathbb{F}),k]_\Theta$.

□

Next we give another characterization of the space $[\Gamma(n;\mathbb{F}),k]_\Theta$.

Definition. A modular form $f \in [\Gamma(n;\mathbb{F}),k]$ is called stable if there exists $N \geq n$ and a singular $F \in [\Gamma(N;\mathbb{F}),k]$ such that $f = F|\phi^{N-n}$.

We observe from Theorem 4.2 that the set of stable modular forms becomes a subspace.

Theorem 4.3. Let $k \equiv 0 \bmod 4$ and $k > 0$. Then the subspace of stable modular forms of weight k coincides with $[\Gamma(n;\mathbb{F}),k]_\Theta$.

Proof. One has $\Theta^{(n)}(\cdot,S;\mathbb{F})|\phi = \Theta^{(n-1)}(\cdot,S;\mathbb{F})$ in view of Theorem 2.6. Hence $[\Gamma(n;\mathbb{F}),k]_\Theta$ consists of stable modular forms. On the other hand, if f is stable use Theorem 4.2 and the description of the action of SIEGEL's ϕ-operator.

□

If $f \in [\Gamma(n;\mathbb{F}),k]$ is a cusp form then $f|_{H(n;\mathbb{R})}$ is a cusp form on SIEGEL's half-space, too. The corresponding fact does not hold if we consider singular modular forms. But the preceding theorem and Corollary 2.8 lead to

Corollary 4.4. Let $k \equiv 0 \bmod 4$ and $f \in [\Gamma(n;\mathbb{F}),k]$ be stable. Then $f|_{H(n;\mathbb{R})}$ is stable, too. In the case $\mathbb{F} = \mathbb{H}$, in addition, $f|_{H(n;\mathbb{C})}$ becomes a stable Hermitian modular form.

Now we are going to show that the condition stated in Theorem 4.2 is also sufficient, i.e. if $k < \frac{1}{2}rn$ then every modular form of weight k is singular. Therefore we have to prove two auxiliary statements.

Lemma 4.5. Let $S \in \mathrm{Pos}(m;\mathbb{F})$ be even and $m \equiv 0 \bmod 2$ for $\mathbb{F} = \mathbb{R}$. Then there exists $q \in \mathbb{N}$ depending only on S such that

$$\Theta_{S^{-1}P,SW(CZ+D)^{-1}}(M\langle Z\rangle,S;\Lambda)$$

$$= \nu(M)\, e^{\pi i\tau(S[W],(CZ+D)^{-1}C)} (\det M\{Z\})^{\frac{1}{2}rm}\, \Theta_{S^{-1}P,SW}(Z,S;\Lambda)$$

holds for all $Z \in H(n;\mathbb{F})$, $P \in \mathrm{Mat}(m,n;\mathcal{O}^{\#})$, $W \in \mathrm{Mat}(m,n;\mathbb{F})_{\mathbb{C}}$, $M = \begin{pmatrix} A & B \\ C & D \end{pmatrix} \in \Gamma(n;\mathbb{F})$ satisfying $M \equiv I \bmod q\mathcal{O}$ and $\Lambda = \mathrm{Mat}(m,n;\mathcal{O})$.

Proof. Apply Proposition 3.5 to $M\langle Z\rangle$ instead of Z, to $M^{-1} = \begin{pmatrix} \bar{D}' & -\bar{B}' \\ -\bar{C}' & \bar{A}' \end{pmatrix}$ instead of M, $S^{-1}P$ instead of P, $Q = 0$ and WZ^{-1} instead of W. Because of $\bar{D}'M\langle Z\rangle-\bar{B}' = Z(CZ+D)^{-1}$ one has

$$\Theta_{S^{-1}P,SW(CZ+D)^{-1}}(M\langle Z\rangle,S;\Lambda)$$

$$= \chi_1\, e^{-\pi i\tau(S[WZ^{-1}],Z(CZ+D)^{-1}D)}\, \Theta_{-PB,S^{-1}PA+WZ^{-1}}(-Z^{-1},S^{-1};\Lambda^{\tau}) ,$$

where χ_1 does not depend on W. Proposition 1.3 and Theorem 2.2 lead to

$$\Theta_{-PB,S^{-1}PA+WZ^{-1}}(-Z^{-1},S^{-1};\Lambda^{\tau})$$

$$= e^{\pi i\tau(S[W],Z^{-1}) + 2\pi i\tau(S^{-1}PA,SW)}\, \Theta_{-PB-SW,S^{-1}PA}(-Z^{-1},S^{-1};\Lambda^{\tau})$$

$$= \chi_2\, e^{\pi i\tau(S[W],Z^{-1}) - 2\pi i\tau(S^{-1}PA,PB)}\, \Theta_{S^{-1}PA,PB+SW}(Z,S;\Lambda)$$

$$= \chi_2\, e^{\pi i\tau(S[W],Z^{-1})}\, \Theta_{S^{-1}PA,SW}(Z,S;\Lambda) ,$$

since PB belongs to Λ^{τ}, where χ_2 does not depend on W.

Because of $(CZ+D)^{-1}DZ^{-1} - Z^{-1} = -(CZ+D)^{-1}C$ Proposition 1.3 yields

$$\Theta_{S^{-1}P,SW(CZ+D)^{-1}}(M\langle Z\rangle,S;\Lambda)$$

$$= \chi_3\, e^{\pi i\tau(S[W],(CZ+D)^{-1}C)}\, \Theta_{S^{-1}P,SW}(Z,S;\Lambda)$$

for all $M \in \Gamma(n;\mathbb{F})$ satisfying $M \equiv I \bmod q\mathcal{O}$, where q is the level of S and χ_3 does not depend on W. Choosing $W = 0$ one has

$$\Theta_{S^{-1}P,0}(M\langle Z\rangle,S;\Lambda) = \nu(M)\, (\det M\{Z\})^{\frac{1}{2}rm}\, \Theta_{S^{-1}P,0}(Z,S;\Lambda)$$

for all M belonging to a suitable principal congruence subgroup according to Theorem 3.8.

□

Lemma 4.6. Let $f \in [\Gamma(n;\mathbb{F}),k]$ and $m \in \mathbb{N}$, $0 < m < n$. Then f possesses a FOURIER-expansion of the form

$$(1) \qquad f(Z) = \sum_{T_2 \in \mathrm{Sym}^{\tau}(m;\mathcal{O}),\, T_2 \geq 0} \varphi_{T_2}(Z_0,Z_1)\, e^{2\pi i \tau(T_2,Z_2)},$$

$$Z = \begin{pmatrix} Z_0 & Z_1 \\ * & Z_2 \end{pmatrix}, \quad Z_2 = Z_2^{(m)} .$$

If T_2 is invertible then

$$\varphi_{T_2}(Z_0,Z_1) = \sum_{H \bmod 2T_2} g_H(Z_0)\, \Theta_{(2T_2)^{-1}\bar{H}',\,2T_2\bar{Z}_1'}(Z_0,2T_2;\Lambda(\mathbb{F}))$$

holds, where $H \bmod 2T_2$ means that $H \in \mathrm{Mat}(n-m,n;\mathcal{O}^{\#})$ runs through a set of representatives of the orbits $\{H+2GT_2 \,;\, G \in \mathrm{Mat}(n-m,m;\mathcal{O})\}$. If $rm \equiv 0 \bmod 2$, then g_H is a modular form of degree $n-m$ and weight $k-\frac{1}{2}rm$ with respect to as suitable congruence subgroup and to the character ν depending only on T_2.

Proof. f possesses a FOURIER-expansion

$$f(Z) = \sum_{T} \alpha(T)\, e^{2\pi i \tau(T,Z)}, \quad Z \in H(n;\mathbb{F}) .$$

Considering a decomposition $Z = \begin{pmatrix} Z_0 & Z_1 \\ * & Z_2 \end{pmatrix}$, $T = \begin{pmatrix} T_0 & T_1 \\ * & T_2 \end{pmatrix}$, $Z_2 = Z_2^{(m)}$, $T_2 = T_2^{(m)}$ we obtain (1) by a rearrangement, where

$$\varphi_{T_2}(Z_0,Z_1) = \sum_{T = \begin{pmatrix} T_0 & T_1 \\ * & T_2 \end{pmatrix}} \alpha(T)\, e^{2\pi i \tau(T_0,Z_0) + 4\pi i \tau(T_1,Z_1)} .$$

Since the FOURIER-series converges locally uniformly, the coefficients $\varphi := \varphi_{T_2}$ are holomorphic on $H(n-m;\mathbb{F}) \times \mathrm{Mat}(n-m,m;\mathbb{F})_{\mathbb{C}}$. Given $G_1 \in \mathrm{Mat}(n-m,m;\mathcal{O})$ one has $S = \begin{pmatrix} 0 & G_1 \\ \bar{G}_1' & 0 \end{pmatrix} \in \mathrm{Sym}(n;\mathcal{O})$. Now $f(Z+S) = f(Z)$ and the uniqueness of the FOURIER-expansion lead to

(2) $$\varphi(Z_o,Z_1+G_1) = \varphi(Z_o,Z_1) .$$

Consider $U = \begin{pmatrix} I & G \\ O & I \end{pmatrix}$, where $G \in Mat(n-m,m;\mathcal{O})$. From

$$\begin{pmatrix} Z_o & Z_1 \\ * & Z_2 \end{pmatrix} [U] = \begin{pmatrix} Z_o & Z_oG+Z_1 \\ * & Z_2+Z_o[G]+\bar{Z}_1'G+\bar{G}'Z_1 \end{pmatrix}$$

and $f(Z[U]) = f(Z)$ we conclude that

(3) $$\varphi(Z_o,Z_1) = \varphi(Z_o,Z_1+Z_oG) \quad e^{2\pi i\tau(T_2,Z_o[G]+\bar{Z}_1'G+\bar{G}'Z_1)} .$$

Because of (2) φ has a FOURIER-expansion of the form

$$\varphi(Z_o,Z_1) = \sum_{H\in\Lambda^\tau} \psi_H(Z_o)\, e^{2\pi i\tau(H,Z_1)} ,$$

where $\Lambda = Mat(n-m,m;\mathcal{O})$. Given $G \in \Lambda$ then (3) leads to

$$\varphi(Z_o,Z_1) = \sum_{H\in\Lambda^\tau} \psi_H(Z_o)\, e^{2\pi i\tau(T_2,Z_o[G])+2\pi i\tau(H,Z_oG)+2\pi i\tau(H+2GT_2,Z_1)} .$$

Because of $2GT_2 \in \Lambda^\tau$ the uniqueness of the FOURIER-coefficients yields

$$\psi_{H+2GT_2}(Z_o) = \psi_H(Z_o)\, e^{2\pi i\tau(T_2,Z_o[G]) + 2\pi i\tau(H,Z_oG)} .$$

Let T_2 be invertible. Then there exist only finitely many orbits $\{H+2GT_2 ; G \in \Lambda\}$, $H \in \Lambda^\tau$. Hence it follows that

$$\varphi(Z_o,Z_1) = \sum_{H \bmod 2T_2} \psi_H(Z_o) \sum_{G\in\Lambda} e^{2\pi i(\tau(T_2,Z_o[G])+\tau(H,Z_oG)+\tau(H+2GT_2,Z_1))} .$$

Proposition 1.1 yields

$$2\tau(T_2,Z_o[G]) + 2\tau(H,Z_oG) + 2\tau(H+2GT_2,Z_1)$$
$$= \tau(2T_2[\bar{G}'+(2T_2)^{-1}\bar{H}'],Z_o)+2\tau(\bar{G}'+(2T_2)^{-1}\bar{H}',2T_2\bar{Z}_1') - \tau((2T_2)^{-1},Z_o[H]) .$$

Defining $g_H(Z_o) := \psi_H(Z_o)\, e^{-\pi i\tau((2T_2)^{-1},Z_o[H])}$ we have

$$\varphi(Z_o,Z_1) = \sum_{H \bmod 2T_2} g_H(Z_o)\ \Theta_{(2T_2)^{-1}\bar{H}',2T_2\bar{Z}_1'}(Z_o,2T_2;\Lambda) .$$

Let $M_o \in \Gamma(n-m;\mathbb{F})$, $M_o \equiv I \bmod q\mathcal{O}$. Consider $M := M_o \times I \in \Gamma(n;\mathbb{F})$. One easily computes that

$$M\langle Z\rangle = \begin{pmatrix} M_o\langle Z_o\rangle & \overline{(C_oZ_o+D_o)}'^{-1}Z_1 \\ * & Z_2-\bar{Z}_1'(C_oZ_o+D_o)^{-1}C_oZ_1 \end{pmatrix} .$$

Since f is a modular form of weight k , one has

$$f(M\langle Z\rangle) = (\det M\{Z\})^k f(Z) ,$$

$$\varphi(M_o\langle Z_o\rangle, \overline{(C_oZ_o+D_o)}'^{-1}Z_1)$$

$$= e^{2\pi i\tau(T_2[\bar{Z}_1'],(C_oZ_o+D_o)^{-1}C_o)} (\det M_o\{Z_o\})^k \varphi(Z_o,Z_1)$$

if we observe that $(\det M\{Z\})^k = (\det M_o\{Z_o\})^k$. Now use the preceding lemma with $S := 2T_2$, $P := \bar{H}'$ and $W := \bar{Z}_1'$. Suppose that $rm \equiv 0 \bmod 2$. Then g_H turns out to be a modular form of weight $k-\frac{1}{2}rm$ with respect to a congruence subgroup and the character υ depending only on T_2 , since the theta-series $\Theta_{P,Q}(Z,S;\Lambda)$ considered as a function of Q does not vanish identically.

□

Thus we have the crucial tool in order to prove

<u>Theorem 4.7.</u> <u>Let</u> $k \in \mathbb{Z}$, $k \equiv 0 \bmod 2$ <u>in the case</u> $\mathbb{F} = \mathbb{H}$, <u>and</u> $k < \frac{1}{2}rn$. <u>Then every</u> $f \in [\Gamma(n;\mathbb{F}),k]$ <u>is singular</u>.

<u>Proof.</u> The assertion is trivial for $n = 1$ or $k \leq 0$. Let $n \geq 2$. If $\mathbb{F} = \mathbb{C},\mathbb{H}$ we apply Lemma 4.6 to the case $m = n-1$. In the case of SIEGEL's modular forms we may provide that $n > 2$. In this case we apply Lemma 4.6 again and put $m = n-1$, if n is odd, and $m = n-2$, if n is even. Thus, in any case, the weight $l = k-\frac{1}{2}rm$ of the forms g_H in Lemma 4.6 is not positive. If l is negative then the forms g_H vanish identically in view of Theorem III.3.2. Hence f possesses a FOURIER-expansion of the form

$$f(Z) = \sum_{\substack{T=\begin{pmatrix} T_o & T_1 \\ * & T_2\end{pmatrix}, \\ \det T_2 = 0}} \alpha(T)\, e^{2\pi i\tau(T,Z)} , \quad Z \in H(n;\mathbb{F}) .$$

It follows that $\alpha(T) = 0$ for all $T > 0$.

Therefore let $l = 0$. Then the forms g_H are constant in view of Theorem III.3.2. Given a positive definite $T_2 \in \mathrm{Sym}^{\tau}(m;\mathcal{O})$ Lemma 4.6 leads to

$$\varphi_{T_2}(Z_o,Z_1) = \sum_{\substack{H \bmod 2T_2 \\ G\in\Lambda}} g_H\, e^{2\pi i\tau(T_2^{-1}[T_2\bar{G}'+\frac{1}{2}\bar{H}'],Z_o)+4\pi i\tau(T_2\bar{G}'+\frac{1}{2}\bar{H}',\bar{Z}_1')} .$$

Defining $R := T_2\bar{G}'+\bar{H}'$ and $T = \begin{pmatrix} T_2^{-1}[R] & \bar{R}' \\ R & T_2 \end{pmatrix}$ one computes that

$$\tau(T_2^{-1}[T_2G'+\tfrac{1}{2}H'],Z_o) + 2\tau(T_2G'+\tfrac{1}{2}\bar{H}',\bar{Z}_1') + \tau(T_2,Z_2) = \tau(T,Z) .$$

Now $\det T = 0$ implies that $\alpha(T) = 0$ for all $T > 0$. Hence f turns out to be singular.

□

Theorem 4.7, Theorem 4.2 and Lemma 2.9 lead to

<u>Corollary 4.8.</u> <u>Let</u> $k \in \mathbb{N}$, $k \equiv 0 \bmod 2$ <u>in the case</u> $\mathbb{F} = \mathbb{H}$ <u>and</u> $k < \frac{1}{2}rn$. <u>Then all modular forms of weight</u> k <u>vanish identically</u>, <u>whenever</u> $k \not\equiv 0 \bmod 4$. <u>In the case</u> $k \equiv 0 \bmod 4$ <u>one has</u>

$$[\Gamma(n;\mathbb{F}),k] = [\Gamma(n;\mathbb{F}),k]_{\Theta} ,$$

$$\dim\,[\Gamma(n;\mathbb{F}),k] = \mathrm{st}(m;\mathbb{F}) ,$$

<u>where</u> $m = \frac{2k}{r}$.

The preceding corollary and Theorem 2.6 yield

<u>Corollary 4.9.</u> <u>Let</u> $k \in \mathbb{N}_o$, $k \equiv 0 \bmod 4$. <u>Then</u> SIEGEL'<u>s</u> ϕ-<u>operator</u>

$$\phi : [\Gamma(n;\mathbb{F}),k] \longrightarrow [\Gamma(n-1;\mathbb{F}),k]$$

<u>is injective</u>, <u>whenever</u> $k < \frac{1}{2}rn$, <u>and bijective</u>, <u>whenever</u> $k < \frac{1}{2}r(n-1)$.

<u>Corollary 4.10.</u> a) $[\Gamma(n;\mathbb{F}),2] = \{0\}$.

b) $$[\Gamma(n;\mathbb{H}),4] = \mathbb{C}\,\Theta^{(n)}(\cdot,S_{\mathbb{H}};\mathbb{H}) ,$$

<u>where</u>

$$S_{\mathbb{H}} = \begin{pmatrix} 2 & e_1+e_2 \\ e_1-e_2 & 2 \end{pmatrix} .$$

<u>Proof.</u> If $\mathbb{F} = \mathbb{C}, \mathbb{H}$ and $n > 2$ Corollary 4.8 leads to the assertion. In the case $n \leqq 2$ confer the schedule in III, §2 and Corollary 2.7.

Let $\mathbb{F} = \mathbb{R}$. If $n > 4$ apply Corollary 4.8. In the case $n \leqq 4$ confer EICHLER [16].

□

Moreover, we conclude from EICHLER [16], Corollary 4.8 and $st(8;\mathbb{R}) = 1$ that

$$[\Gamma(n;\mathbb{R}),4] = \mathbb{C}\,\Theta^{(n)}(\cdot,S_{\mathbb{R}};\mathbb{R}) \ , \text{ if } n \leqq 4 \text{ or } n \geqq 9 \ .$$

It follows from Corollary 4.9 that SIEGEL's ϕ-operator turns out to be bijective, whenever the weight is sufficiently small. On the other hand, it will be shown in the next chapter by means of EISENSTEIN-series that SIEGEL's ϕ-operator becomes surjective, whenever the weight is sufficiently large in proportion to the degree.

Chapter V EISENSTEIN- and POINCARE´-series

If the weight is "small" in proportion to the degree, the vector space of modular forms is generated by theta-series. EISENSTEIN- and POINCARE´-series form a set of generators, whenever the weight is sufficiently "large" in proportion to the degree. Moreover, a certain type of POINCARE´-series is particularly suitable for the construction of algebraically independent modular forms.

§1 deals with auxiliary computations. Some integrals, which will be used later on, are estimated. Especially, the generalized beta- and gamma-integral are calculated.

In §2 we introduce another method of constructing non-identically vanishing modular forms which are given by EISENSTEIN-series. This general type of EISENSTEIN-series was investigated by KLINGEN [32]. In particular, the behavior of convergence is examined carefully.

In §3 we introduce two different kinds of POINCARE´-series investigated by MAASS [43] resp. KLINGEN [31]. All these POINCARE´-series can be considered as special EISENSTEIN-series.

§4 deals with a generalization of PETERSSON's metrization principle. By means of the introduced scalar product we can derive theorems of representation saying that the vector space of modular forms is generated by certain types of EISENSTEIN- resp. POINCARE´-series, whenever the weight is sufficiently "large".

In §5 we separate inequivalent points of the half-space by cusp forms. Moreover, we construct $h+1$ algebraically independent cusp forms given by POINCARE´-series.

§1 Integrals

In this paragraph we compute some integrals, which appear in the proofs of convergence later on.

According to SIEGEL [53] we estimate the generalized beta-integral. Let

$$\Gamma(t) := \int_0^\infty x^{t-1} e^{-x} dx \ , \ t > 0 \ ,$$

denote the gamma-function and $h = \dim \mathrm{Sym}(n;\mathbb{F}) = n + \frac{1}{2}rn(n-1)$.

Lemma 1.1. *Defining*

$$\beta(n,s,t) = \int_{\mathrm{Pos}(n;\mathbb{F})} (\det Y)^{s-h/n} (\det(Y+I))^{-t} dY$$

then

$$\beta(n,s,t) = \pi^{\frac{1}{4}rn(n-1)} \prod_{k=0}^{n-1} \frac{\Gamma(s-\frac{1}{2}rk)\Gamma(t-s-\frac{1}{2}rk)}{\Gamma(t-\frac{1}{2}rk)}$$

holds, whenever s,t *belong to* $\mathbb{R}$ *and satisfy* $s > \frac{1}{2}r(n-1)$ *as well as* $t-s > \frac{1}{2}r(n-1)$.

Proof. We use induction on n . Let $n = 1$. It follows from the definition of the gamma-function that

$$(1) \qquad \Gamma(t)(1+y)^{-t} = \int_0^\infty x^{t-1} e^{-(1+y)x} dx$$

holds for $y > 0$ and $t > 0$. Given $s > 0$ and $t > s$ then (1) leads to

$$\begin{aligned} &\Gamma(t)\ \beta(1,s,t) \\ &= \int_0^\infty y^{s-1} \int_0^\infty x^{t-1} e^{-(1+y)x} dx\, dy \\ &= \int_0^\infty x^{t-1} e^{-x} \int_0^\infty y^{s-1} e^{-yx} dy\, dx \\ &= \Gamma(s) \int_0^\infty x^{t-s-1} e^{-x} dx \\ &= \Gamma(s)\ \Gamma(t-s) \ . \end{aligned}$$

Now let $n > 1$, $s > \frac{1}{2}r(n-1)$, $t-s > \frac{1}{2}r(n-1)$. Consider

$$Y = \begin{pmatrix} y & \bar{\mathfrak{y}}' \\ \mathfrak{y} & Y_1 \end{pmatrix} \in \mathrm{Pos}(n;\mathbb{F}) \quad ,$$

where $\mathfrak{y} \in \mathbb{F}^{n-1}$, $Y_1 \in \mathrm{Pos}(n-1;\mathbb{F})$ and $y > 0$. Lemma I.3.2 yields

$$\det Y = (y - Y_1^{-1}[\mathfrak{y}]) \det Y_1 \; ,$$

$$\det(I+Y) = (1 + y - (I+Y_1)^{-1}[\mathfrak{y}]) \det(I+Y_1) \; .$$

Defining

$$a(Y_1) = \int_{\mathfrak{y}\in\mathbb{F}^{n-1}} \int_{y>Y_1^{-1}[\mathfrak{y}]} (y-Y_1^{-1}[\mathfrak{y}])^{s-h/n} (1 + y-(I+Y_1)^{-1}[\mathfrak{y}])^{-t} \, dy \, d\mathfrak{y}$$

$$= \int_{\mathfrak{y}\in\mathbb{F}^{n-1}} \int_{z>0} z^{s-h/n} \left(1+z+(Y_1^{-1}-(Y_1+I)^{-1})[\mathfrak{y}]\right)^{-t} dz \, d\mathfrak{y}$$

one has

(2) $$\beta(n,s,t) = \int_{\mathrm{Pos}(n-1;\mathbb{F})} a(Y_1) \, (\det Y_1)^{s-h/n} \, (\det(Y_1+I))^{-t} \, dY_1 \; .$$

Because of $Y_1^{-1} > (Y_1+I)^{-1}$ there exists $U \in GL(n-1;\mathbb{F})$ satisfying $Y_1^{-1} - (Y_1+I)^{-1} = \bar{U}'U$, hence

$$\det \bar{U}'U = \det(Y_1^{-1} - (Y_1+I)^{-1}) = (\det Y_1)^{-1} \, (\det(Y_1+I))^{-1} \; .$$

The absolute value of the JACOBIan determinant of the transformation

$$\mathbb{F}^{n-1} \longrightarrow \mathbb{F}^{n-1} \; , \; \mathfrak{y} \longmapsto \mathfrak{z} := U\mathfrak{y} \; ,$$

is given by

$$|\det \hat{U}| = (\det Y_1)^{-\frac{1}{2}r} \, (\det(Y_1+I))^{-\frac{1}{2}r} =: c^{-1} \; .$$

Change of variables leads to

$$a(Y_1) = c \int_{\mathfrak{z}\in\mathbb{F}^{n-1}} \int_{z>0} z^{s-h/n} \, (1+z+\bar{\mathfrak{z}}'\mathfrak{z})^{-t} \, dz \, d\mathfrak{z} \; .$$

Regarding (1) it follows that

$$\Gamma(t) \, (1+z+\bar{\mathfrak{z}}'\mathfrak{z})^{-t} = \int_{x>0} x^{t-1} \, e^{-(1+z+\bar{\mathfrak{z}}'\mathfrak{z})x} \, dx \; ,$$

$$\Gamma(t) \, a(Y_1)$$

$$= c \int_{\mathfrak{z}\in\mathbb{F}^{n-1}} \int_{z>0} \int_{x>0} z^{s-h/n} \, x^{t-1} \, e^{-(1+z+\bar{\mathfrak{z}}'\mathfrak{z})x} \, dx \, dz \, d\mathfrak{z}$$

$$= c \int_{\mathfrak{z}\in\mathbb{F}^{n-1}} \int_{x>0} x^{t-1} \, e^{-(1+\bar{\mathfrak{z}}'\mathfrak{z})x} \int_{z>0} z^{s-\frac{1}{2}r(n-1)-1} e^{-zx} \, dz \, dx \, d\mathfrak{z}$$

$$= c\ \Gamma(s-\tfrac{1}{2}r(n-1)) \int_{x>0} x^{t-s+\frac{1}{2}r(n-1)-1}\, e^{-x} \int_{\mathfrak{z}\in\mathbb{F}^{n-1}} e^{-\bar{\mathfrak{z}}'\mathfrak{z}x}\, d\mathfrak{z}\, dx.$$

Given $v > 0$ one has $\int_{-\infty}^{+\infty} e^{-u^2 v}\, du = \pi^{\frac{1}{2}} v^{-\frac{1}{2}}$, hence

$$\int_{\mathfrak{z}\in\mathbb{F}^{n-1}} e^{-\bar{\mathfrak{z}}'\mathfrak{z}x}\, d\mathfrak{z} = \pi^{\frac{1}{2}r(n-1)}\, x^{-\frac{1}{2}r(n-1)}$$

holds for all $x > 0$. Thus

$$\Gamma(t)\ a(Y_1) = \pi^{\frac{1}{2}r(n-1)}\ \Gamma(s-\tfrac{1}{2}r(n-1))\ c \int_{x>0} x^{t-s-1}\, e^{-x}\, dx$$

$$= \pi^{\frac{1}{2}r(n-1)}\ \Gamma(s-\tfrac{1}{2}r(n-1))\ \Gamma(t-s)\ c\ .$$

Using this identity (2) leads to

$$\beta(n,s,t) = \pi^{\frac{1}{2}r(n-1)}\ \frac{\Gamma(s-\frac{1}{2}r(n-1))\ \Gamma(t-s)}{\Gamma(t)}\ \beta(n-1,s,t-\tfrac{1}{2}r)\ .$$

The induction hypothesis completes the proof.

□

According to BRAUN [6] we prove

<u>Lemma 1.2.</u> <u>Given</u> $Z = X + iY \in H(n;\mathbb{F})$ <u>and real</u> $k > 1+r(n-1)$ <u>the integral</u>

$$\eta_k(Z) := \int_{Sym(n;\mathbb{F})} |\det(Z+S)|^{-k}\, dS$$

<u>exists and fulfills</u>

$$\eta_k(Z) = (\det Y)^{-k+h/n}\ \eta_k(iI)\ .$$

<u>Proof.</u> Change of variables leads to the second assertion. Hence it suffices to demonstrate the convergence of $\eta_{k,n} := \eta_k(iI^{(n)})$ by induction on n . If $n = 1$ and $k > 1$, one has in view of (1)

$$\eta_{k,1} = \int_{-\infty}^{+\infty} (1+s^2)^{-\frac{1}{2}k}\, ds\ ,$$

$$\Gamma(\tfrac{1}{2}k)\ \eta_{k,1} = \int_{-\infty}^{+\infty}\int_{0}^{\infty} x^{\frac{1}{2}k-1}\, e^{-(1+s^2)x}\, dx\, ds$$

$$= \int_{0}^{\infty} x^{\frac{1}{2}k-1}\, e^{-x} \int_{-\infty}^{+\infty} e^{-s^2 x}\, ds\, dx$$

$$= \pi^{\frac{1}{2}} \int_0^\infty x^{\frac{1}{2}(k-1)-1} e^{-x} dx$$

$$= \pi^{\frac{1}{2}} \Gamma(\tfrac{1}{2}(k-1)) .$$

Hence $\eta_{k,1}$ exists, whenever $k > 1$.

Let $n \geq 2$ and consider a decomposition

$$S = \begin{pmatrix} s & \bar{\mathfrak{s}}' \\ \mathfrak{s} & S_1 \end{pmatrix}, \; s \in \mathbb{R}, \; \mathfrak{s} \in \mathbb{F}^{n-1}, \; S_1 \in \mathrm{Sym}(n-1;\mathbb{F}) .$$

Lemma I.3.2 yields

$$\det(S+iI) = (s + i - (S_1+iI)^{-1}[\mathfrak{s}]) \det(S_1+iI) .$$

Put $\mathbf{a}(\mathfrak{s},S_1) := i - (S_1+iI)^{-1}[\mathfrak{s}]$ and

$$a(S_1) := \int_{s\in\mathbb{R}} \int_{\mathfrak{s}\in\mathbb{F}^{n-1}} |s + a(\mathfrak{s},S_1)|^{-k} ds\, d\mathfrak{s} .$$

Change of variables leads to

$$a(S_1) = \int_{\mathfrak{s}\in\mathbb{F}^{n-1}} y^{1-k} \int_{t\in\mathbb{R}} |t+i|^{-k} dt\, d\mathfrak{s} ,$$

where $y := \mathrm{Im}\, a(\mathfrak{s},S_1)$. One has $Y := \mathrm{Im}(-(S_1+iI)^{-1}) = (I+S_1{}^2)^{-1}$ according to Theorem II.1.7 and $y = 1 + Y[\mathfrak{s}]$, hence

$$a(S_1) = |\det(S_1+iI)|^r \int_{\mathfrak{t}\in\mathbb{F}^{n-1}} \int_{t\in\mathbb{R}} |t+i|^{-k} (1+\bar{\mathfrak{t}}'\mathfrak{t})^{1-k} dt\, d\mathfrak{t} .$$

The integral $\int_{-\infty}^{+\infty} |t+i|^{-k} dt$ exists because of the case $n = 1$. We obtain

$$\int_{\mathfrak{t}\in\mathbb{F}^{n-1}} (1 + \bar{\mathfrak{t}}'\mathfrak{t})^{1-k} d\mathfrak{t} = \pi^{\frac{1}{2}r(n-1)} \frac{\Gamma(k-\frac{1}{2}r(n-1)-1)}{\Gamma(k-1)}$$

from the proof of the preceding lemma. Thus it suffices to demonstrate the existence of the integral

$$\int_{S_1\in\mathrm{Sym}(n-1;\mathbb{F})} |\det(S_1+iI)|^{r-k} dS_1 = \eta_{k-r,n-1} .$$

The induction hypothesis completes the proof.

□

These two lemmata enable us to compute an integral, which was estimated by KLINGEN [33] in the case of SIEGEL's half-space. Let dv denote the symplectic volume element (cf. Theorem II.1.10).

<u>Corollary 1.3.</u> <u>Given</u> $k > 1 + r(n-1)$ <u>there exists a constant</u> $\alpha = \alpha(n,k,\mathbb{F})$, <u>such that</u>

$$\int_{H(n;\mathbb{F})} (\det \operatorname{Im} W)^k \, |\det(W+Z)|^{-2k} \, dv(W) = \alpha (\det Y)^{-k}$$

<u>holds for all</u> $Z = X + iY \in H(n;\mathbb{F})$.

<u>Proof.</u> The integral is denoted by $\sigma(Z)$. Change of variables leads to

$$\sigma(Z) = (\det Y)^{-k} \sigma(iI) .$$

Put $W = U + iV$, then Lemma 1.2 yields

$$\int_{Sym(n;\mathbb{F})} |\det(U+i(V+I))|^{-2k} \, dU = (\det(V+I))^{-2k+h/n} \, \eta_{2k}(iI) ,$$

since $2k > 1 + r(n-1)$. Because of $dv = (\det V)^{-2h/n} dU\,dV$ it suffices to estimate

$$\int_{Pos(n;\mathbb{F})} (\det V)^{k-2h/n} (\det(V+I))^{-2k+h/n} \, dV .$$

This integral converges according to Lemma 1.1.

□

We want to demonstrate the convergence of a series by the convergence of the integral in Lemma 1.2. Therefore we need

<u>Proposition 1.4.</u> <u>Given</u> $\varepsilon > 0$ <u>then</u>

$$c^{-1} |\det Z| \le |\det(Z+S)| \le c \, |\det Z|$$

<u>holds for all</u> $Z \in H(n;\mathbb{F})$ <u>and all</u> $S \in Sym(n;\mathbb{F})$ <u>satisfying</u> $S^2 \le \varepsilon^{-2} I$, <u>where</u> $c = 2^{\frac{1}{2}n}(1 + \varepsilon^{-1} tr(Y^{-1}))^n$.

<u>Proof.</u> Theorem II.1.7 yields $|\det Z|^2 = \det Y \, \det(Y+Y^{-1}[X])$. Because of $Y^{-1} \le tr(Y^{-1}) I$ one has

$$\begin{aligned} Y + Y^{-1}[X+S] &\le Y + 2Y^{-1}[X] + 2Y^{-1}[S] \\ &\le Y + 2Y^{-1}[X] + 2tr(Y^{-1})\varepsilon^{-2} I \\ &\le 2(1 + \varepsilon^{-2}(tr(Y^{-1}))^2) \, (Y + Y^{-1}[X]) . \end{aligned}$$

Observing $(1 + \varepsilon^{-2}(tr(Y^{-1}))^2)^{\frac{1}{2}n} \le (1 + \varepsilon^{-1} tr(Y^{-1}))^n$ leads to

$$|\det(Z+S)| \le c \, |\det Z| .$$

Replacing Z by $Z+S$ and S by $-S$ completes the proof.

□

Thus we are able to prove

Corollary 1.5. Given $k > 1 + r(n-1)$ the series

$$\sum_{S \in Sym(n;\mathcal{O})} |\det(Z+S)|^{-k}, \quad Z \in H(n;\mathbb{F}),$$

converges. More precisely, given $\varepsilon > 0$ there exists a constant $c = c(n,\varepsilon) > 0$, such that

$$c^{-k}\, \eta_k(Z) \leq \sum_{S \in Sym(n;\mathcal{O})} |\det(Z+S)|^{-k} \leq c^k\, \eta_k(Z)$$

holds for all $Z = X + iY \in H(n;\mathbb{F})$ satisfying $Y \geq \varepsilon I$.

Proof. Let $C(n;\mathbb{F})$ denote a fundamental parallelotope of the lattice $Sym(n;\mathcal{O})$ in $Sym(n;\mathbb{F})$. We obtain the assertion by an estimation between $|\det(Z+S)|^{-k}$ and

$$\int_{C(n;\mathbb{F})} |\det(Z+S+H)|^{-k}\, dH$$

in view of Proposition 1.4.

□

According to SIEGEL [53], Hilfssatz 37, we compute the generalized gamma-integral.

Lemma 1.6. Given $k \in \mathbb{N}$, $k > \frac{1}{2}r(n-1)$ and $Z \in H(n;\mathbb{F})$ one has

$$\int_{Pos(n;\mathbb{F})} (\det S)^{k-h/n}\, e^{2\pi i \tau(S,Z)}\, dS$$

$$= (\det \tfrac{1}{i}Z)^{-k}\, \pi^{\frac{1}{4}rn(n-1)}\, (2\pi)^{-nk} \prod_{l=0}^{n-1} \Gamma(k-\tfrac{1}{2}rl) .$$

Proof. Since both sides represent holomorphic functions, we can restrict ourselves to the case $Z = iY$. Changing variables we may assume that $Z = iI$ and compute the integral

$$\gamma(n,k) := \int_{Pos(n;\mathbb{F})} (\det S)^{k-h/n} e^{-2\pi tr(S)} dS$$

by induction on n. In the case $n = 1$ one has $k > 0$ and (1) leads to

$$\gamma(1,k) = \int_0^\infty s^{k-1} e^{-2\pi s} ds = (2\pi)^{-k} \Gamma(k) .$$

Let $n \geq 2$. Given $S \in Pos(n;\mathbb{F})$ we choose a decomposition

$$S = \begin{pmatrix} s & \bar{\mathfrak{s}}' \\ \mathfrak{s} & S_1 \end{pmatrix}, \text{ where } s > 0 , \mathfrak{s} \in \mathbb{F}^{n-1} , S_1 \in Pos(n-1;\mathbb{F}) .$$

Because of $\det S = (s - S_1^{-1}[\mathfrak{s}]) \det S_1$ one has

$$\gamma(n,k) = \int_{Pos(n-1;\mathbb{F})} a(S_1)(\det S_1)^{k-h/n} e^{-2\pi tr(S_1)} dS_1 ,$$

where

$$a(S_1) = \int_{\mathfrak{s}\in\mathbb{F}^{n-1}} \int_{s>S_1^{-1}[\mathfrak{s}]} (s - S_1^{-1}[\mathfrak{s}])^{k-h/n} e^{-2\pi s} ds\, d\mathfrak{s}$$

$$= \int_{\mathfrak{s}\in\mathbb{F}^{n-1}} \int_{t>0} t^{k-h/n} e^{-2\pi t-2\pi S_1^{-1}[\mathfrak{s}]} dt\, d\mathfrak{s} .$$

Now (1) leads to

$$a(S_1) = \Gamma(k - \tfrac{1}{2}r(n-1)) \ (2\pi)^{-k+\frac{1}{2}r(n-1)} \int_{\mathfrak{s}\in\mathbb{F}^{n-1}} e^{-2\pi S_1^{-1}[\mathfrak{s}]} d\mathfrak{s}$$

$$= 2^{-\frac{1}{2}r(n-1)} (\det S_1)^{\frac{1}{2}r} (2\pi)^{-k+\frac{1}{2}r(n-1)} \Gamma(k-\tfrac{1}{2}r(n-1)) .$$

Thus we obtain

$$\gamma(n,k) = (2\pi)^{-k} \pi^{\frac{1}{2}r(n-1)} \Gamma(k-\tfrac{1}{2}r(n-1)) \gamma(n-1,k) .$$

The induction hypothesis completes the proof.

□

Finally we use the preceding lemma to derive an identity between two infinite series.

<u>Lemma 1.7.</u> <u>Given</u> $k \in \mathbb{N}$, $k > 1 + r(n-1)$ <u>and</u> $Z \in H(n;\mathbb{F})$ <u>one has</u>

$$\sum_{T\in Sym^{\tau}(n;\mathcal{O}),T>0} (\det T)^{k-h/n} e^{2\pi i\tau(T,Z)}$$

$$= \delta(n,k,\mathcal{O}) \sum_{S\in Sym(n;\mathcal{O})} \det(Z+S)^{-k} ,$$

<u>where</u> $\delta(n,k,\mathcal{O}) = \text{vol Sym}(n;\mathcal{O})\ (4\pi)^{\frac{1}{4}rn(n-1)}\ (-2\pi i)^{-nk} \prod_{l=0}^{n-1} \Gamma(k-\frac{1}{2}rl)$.

<u>Proof.</u> The first series converges absolutely according to Proposition III.1.3. Given $x = (x_{11},\dots,x_{nn},x_{12}^{(1)},\dots,x_{12}^{(r)},\dots,x_{n-1n}^{(r)})' \in \mathbb{R}^h$ we define $\varphi(x) := X = (x_{kl}) \in \text{Sym}(n;\mathbb{F})$, where $x_{kl} = \sum_{j=1}^{r} x_{kl}^{(j)} e_j$, $1 \le k < l \le n$. Let $A = [I^{(n)},\frac{1}{2}a,\dots,\frac{1}{2}a] \in \text{Mat}(h;\mathbb{R})$ and $B = [I^{(n)},b,\dots,b] \in \text{Mat}(h;\mathbb{R})$, where $b = (a')^{-1}$, $a = a(\mathbb{F}) \in \text{Mat}(r;\mathbb{R})$, $a(\mathbb{F}) = I^{(r)}$, if $\mathbb{F} = \mathbb{R},\mathbb{C}$, and

$$a(\mathbb{H}) = \begin{pmatrix} 1 & 0 & 0 & -1 \\ 0 & 1 & 0 & -1 \\ 0 & 0 & 1 & -1 \\ 0 & 0 & 0 & 2 \end{pmatrix} .$$

The maps $\mathbb{Z}^h \longrightarrow \text{Sym}^\tau(n;\mathcal{O})$, $\mathfrak{g} \longmapsto \varphi(A\mathfrak{g})$, and $\mathbb{Z}^h \longrightarrow \text{Sym}(n;\mathcal{O})$, $\mathfrak{g} \longmapsto \varphi(B\mathfrak{g})$, are bijective. Put

$$\psi : \mathbb{R}^h \longrightarrow \mathbb{C} ,$$

$$\mathfrak{z} \longmapsto \begin{cases} (\det \varphi(A\mathfrak{z}))^{k-h/n}\, e^{2\pi i \tau(\varphi(A\mathfrak{z}),Z)} , & \text{if } \varphi(A\mathfrak{z}) > 0 , \\ 0 & , \text{else} . \end{cases}$$

ψ turns out to be continuous. The application of the POISSON summation formula (cf. [58],VII.2.6) yields

$$\sum_{T\in\text{Sym}^\tau(n;\mathcal{O}),T>0} (\det T)^{k-h/n}\, e^{2\pi i \tau(T,Z)}$$

$$= \sum_{\mathfrak{g}\in\mathbb{Z}^h} \psi(\mathfrak{g}) = \sum_{\mathfrak{g}\in\mathbb{Z}^h} \int_{\mathbb{R}^h} \psi(\mathfrak{z})\, e^{-2\pi i \mathfrak{g}'\mathfrak{z}}\, d\mathfrak{z} .$$

Change of variables leads to

$$(\det A) \int_{\mathbb{R}^h} \psi(\mathfrak{z})\, e^{-2\pi i \mathfrak{g}'\mathfrak{z}}\, d\mathfrak{z}$$

$$= \int_{\mathfrak{y}\in\mathbb{R}^h,\varphi(\mathfrak{y})>0} (\det \varphi(\mathfrak{y}))^{k-h/n}\, e^{2\pi i \tau(\varphi(\mathfrak{y}),Z) - 2\pi i (A'^{-1}\mathfrak{g})'\mathfrak{y}}\, d\mathfrak{y}$$

$$= \int_{\text{Pos}(n;\mathbb{F})} (\det P)^{k-h/n}\, e^{2\pi i \tau(P,Z-S)}\, dP ,$$

where $S := \varphi(B\mathfrak{g})$. Lemma 1.6 completes the proof. □

§2 EISENSTEIN-series

EISENSTEIN-series represent another type of non-identically vanishing modular forms. The classical EISENSTEIN-series were generalized by SIEGEL [53], who introduced analogously built series on $H(n;\mathbb{R})$. In this paragraph we examine another kind of series investigated by KLINGEN [32]. It is essential to survey the behavior of convergence of these series. EISENSTEIN-series turn out to converge absolutely and uniformly in relatively "large" domains, the so-called vertical strips. The representation is based on [32] and [21],I,§5 .

First we have to examine certain subgroups of the symplectic group. Let $\Sigma_n := Sp(n;\mathbb{F})$. Given $0 \leq j \leq n$ we define the set

$$\Sigma_{n,j} := \left\{ M = \begin{pmatrix} * & * \\ 0 & * \end{pmatrix} \in \Sigma_n \ ; \ 0 = 0^{(n-j,n+j)} \right\}$$

consisting of all those symplectic matrices, where the elements in the first $n+j$ columns and the last $n-j$ rows vanish. Obviously, $\Sigma_{n,n} = \Sigma_n$ holds. $\Sigma_{n,o}$ was examined in Proposition II.1.2. Given $M \in \Sigma_{n,j}$ we always choose a decomposition into blocks

$$(1) \qquad M = \begin{pmatrix} A & B \\ C & D \end{pmatrix}, \quad A = \begin{pmatrix} A_1 & A_2 \\ A_3 & A_4 \end{pmatrix}, \quad B = \begin{pmatrix} B_1 & B_2 \\ B_3 & B_4 \end{pmatrix} \quad \text{etc.},$$

where $A_1 = A_1^{(j)}$, $B_1 = B_1^{(j)}$. Indeed, the cases $j = 0,n$ can be interpreted unmistakably.

<u>Proposition 2.1.</u> a) <u>Given</u> $M \in \Sigma_{n,j}$ <u>of the form</u> (1) <u>one has</u> $C_3 = D_3 = 0$, $A_2 = C_2 = 0$ <u>and</u> $C_4 = 0$. $\Sigma_{n,j}$ <u>forms a subgroup of</u> Σ_n .

b) <u>If</u> $j > 0$, <u>then</u> $M_1 := \begin{pmatrix} A_1 & B_1 \\ C_1 & D_1 \end{pmatrix}$ <u>belongs to</u> Σ_j <u>and</u>

$$\chi : \Sigma_{n,j} \longrightarrow \Sigma_j \ , \ M \longmapsto M_1 \ ,$$

<u>becomes a surjective homomorphism of the groups.</u>

c) <u>Let</u> $j > 0$, $Z = \begin{pmatrix} Z_1 & * \\ * & * \end{pmatrix} \in H(n;\mathbb{F})$, <u>where</u> $Z_1 \in H(j;\mathbb{F})$ <u>and</u> $M \in \Sigma_{n,j}$, <u>then</u>

$$M\langle Z\rangle = \begin{pmatrix} M_1\langle Z_1\rangle & * \\ * & * \end{pmatrix} .$$

Proof. We conclude from Lemma II.1.1 that every $M \in \Sigma_{n,j}$ has the form

$$(2) \qquad M = \begin{pmatrix} A_1 & O & B_1 & B_2 \\ A_3 & A_4 & B_3 & B_4 \\ C_1 & O & D_1 & D_2 \\ O & O & O & D_4 \end{pmatrix}$$

Now straightforward computations complete the proof. □

Throughout §2 to §4 let the lower index 1 denote the left upper j-rowed block of an n-rowed matrix, i.e.

$$Z_1 = Z\begin{bmatrix} I \\ O \end{bmatrix}, \text{ where } I = I^{(j)} .$$

Obviously,

$$\Gamma_{n,j} := \Sigma_{n,j} \cap \mathrm{Mat}(2n;\mathcal{O})$$

turns out to be a subgroup of the modular group and

$$\chi : \Gamma_{n,j} \longrightarrow \Gamma_j$$

becomes a surjective homomorphism of the groups, whenever $j > 0$. Especially, one has $\Gamma_{n,n} = \Gamma(n;\mathbb{F})$ and Proposition II.1.2 leads to

$$\Gamma_{n,o} = \left\{ \begin{pmatrix} \bar{U}' & \bar{U}'S \\ O & U^{-1} \end{pmatrix} ; U \in GL(n;\mathcal{O}) , S \in \mathrm{Sym}(n;\mathcal{O}) \right\} .$$

We are going to describe a fundamental domain of the half-space with respect to the action of $\Gamma_{n,j}$. Given $m,n \in \mathbb{N}$ let $\mathcal{C}(m,n;\mathbb{F})$ denote a fundamental parallelotope of the lattice $\mathrm{Mat}(m,n;\mathcal{O})$ in $\mathrm{Mat}(m,n;\mathbb{F})$ and $\mathcal{C}(n;\mathbb{F})$ of $\mathrm{Sym}(n;\mathcal{O})$ in $\mathrm{Sym}(n;\mathbb{F})$.

Proposition 2.2. There exists a fundamental domain of $H(n;\mathbb{F})$ with respect to the action of $\Gamma_{n,j}$, which is contained in

$$\mathcal{F}_{n,j} := \left\{ Z = X + i\begin{pmatrix} Y_1 & O \\ O & Y_2 \end{pmatrix}\begin{bmatrix} I & Y_3 \\ O & I \end{bmatrix} ; \begin{matrix} Z_1 \in \mathcal{F}(j;\mathbb{F}) , Y_2 \in \mathcal{R}(n-j;\mathbb{F}) \\ Y_3 \in \mathcal{C}(j,n-j;\mathbb{F}) , X \in \mathcal{C}(n;\mathbb{F}) \end{matrix} \right\},$$

where, of course, $\mathcal{F}_{n,o} = \mathcal{C}(n;\mathbb{F}) + i\mathcal{R}(n;\mathbb{F})$ and $\mathcal{F}_{n,n} = \mathcal{F}(n;\mathbb{F})$.

Proof. Without restriction we may suppose that $0 < j < n$. Given $Z \in H(n;\mathbb{F})$ it suffices to find $M \in \Gamma_{n,j}$ satisfying $M\langle Z\rangle \in \mathcal{F}_{n,j}$.

First choose $K \in \Gamma_j$ such that $K\langle Z_1\rangle \in \mathcal{F}(j;\mathbb{F})$. In view of Proposition 2.1 we map provide that $K = I$. Then consider

$$M = \begin{pmatrix} \bar{U}' & SU^{-1} \\ 0 & U^{-1} \end{pmatrix} \in \Gamma_{n,j} ,$$

where $S \in \mathrm{Sym}(n;\mathcal{O})$, $S_1 = 0$ and $U = \begin{pmatrix} I & U_2 \\ 0 & U_4 \end{pmatrix} \in GL(n;\mathcal{O})$. Because of $M_1 = I$ the condition $Z_1 \in \mathcal{F}(j;\mathbb{F})$ remains valid. Given U arbitrarily we can always find S such that $X[U] + S \in \mathcal{C}(n;\mathbb{F})$. Since Y belongs to $\mathrm{Pos}(n;\mathbb{F})$, Lemma I.3.2 yields a representation

$$Y = \begin{pmatrix} Y_1 & 0 \\ 0 & Y_2 \end{pmatrix} \begin{bmatrix} I & Y_3 \\ 0 & I \end{bmatrix} ,$$

where $Y_1 \in \mathrm{Pos}(j;\mathbb{F})$, $Y_2 \in \mathrm{Pos}(n-j;\mathbb{F})$ and $Y_3 \in \mathrm{Mat}(j,n-j;\mathbb{F})$. In view of

$$Y[U] = \begin{pmatrix} Y_1 & 0 \\ 0 & Y_2[U_4] \end{pmatrix} \begin{bmatrix} I & U_2 + Y_3U_4 \\ 0 & I \end{bmatrix}$$

the assertion is obvious.

□

We need some estimations of determinants.

<u>Proposition 2.3.</u> <u>Given</u> $Z \in H(n;\mathbb{F})$ <u>and</u> $V \in \mathrm{Sym}(n;\mathbb{F})$, $V \geq 0$, <u>then</u>

$$|\det Z| \leq |\det(Z+iV)| ,$$

<u>where equality holds if and only if</u> $V = 0$.

<u>Proof.</u> Given $Z \in H(n;\mathbb{F})$ one has $\hat{Z} \in H(rn;\mathbb{R})$ and $\det \hat{Z} = (\det Z)^r$. In the case $\mathbb{F} = \mathbb{R}$ we obtain the assertion from CHRISTIAN [13],4.69 .

□

<u>Proposition 2.4.</u> <u>Let</u> $Z \in H(n;\mathbb{F})$ <u>and</u> $W = U + iV \in \mathrm{Sym}(n;\mathbb{F}) \otimes_{\mathbb{R}} \mathbb{C}$, <u>where</u> $V \geq 0$. <u>Defining</u> $a = 2^{-\frac{1}{2}n}\left(1 + \sqrt{\mathrm{tr}(U^2)}\ \mathrm{tr}(Y^{-1})\right)^{-n}$ <u>and</u> $b = a^{-1}\left(1 + \mathrm{tr}(V)\ \mathrm{tr}(Y^{-1})\right)^n$ <u>one has</u>

$$a\,|\det Z| \leq |\det(Z+W)| \leq b\,|\det Z| .$$

<u>Proof.</u> In view of Proposition 2.3 and Proposition 1.4 it suffices to prove

$$|\det(Z+iV)| \leq (1 + \mathrm{tr}(V)\ \mathrm{tr}(Y^{-1}))^n\ |\det Z| .$$

Because of $Y > 0$ and $V \geq 0$ one has $\det(Y+V) \leq \left(1+\mathrm{tr}(V)\mathrm{tr}(Y^{-1})\right)^n \det Y$. Finally

$$\begin{aligned}|\det(Z+iV)|^2 &= \det(Y+V)\ \det(Y+V+(Y+V)^{-1}[X])\\ &\leq \left(1+\mathrm{tr}(V)\ \mathrm{tr}(Y^{-1})\right)^{2n} \det Y \det(Y+Y^{-1}[X])\\ &= \left(1+\mathrm{tr}(V)\ \mathrm{tr}(Y^{-1})\right)^{2n} |\det Z|^2\end{aligned}$$

yields the assertion. □

<u>Definition.</u> Given $\varepsilon > 0$ the set

$$V_\varepsilon(n;\mathbb{F}) := \{Z = X+iY \in H(n;\mathbb{F})\ ;\ Y \geq \varepsilon I\ ,\ X^2 \leq \varepsilon^{-2} I\}$$

is called a <u>vertical strip</u>.

Obviously

$$\hat{V}_\varepsilon(n;\mathbb{F}) := \{\hat{Z}\ ;\ Z \in V_\varepsilon(n;\mathbb{F})\} \subset V_\varepsilon(rn;\mathbb{R}) .$$

Now we are able to give a direct proof of

<u>Lemma 2.5.</u> <u>Given</u> $\varepsilon > 0$ <u>there exists</u> $c = c(n,\varepsilon) > 0$ <u>such that the following holds</u>:

a) $$|\det(Z+W)| \geq c\ |\det(iI+W)|$$

<u>for all</u> $Z \in V_\varepsilon(n;\mathbb{F})$ <u>and</u> $W = U+iV \in \mathrm{Sym}(n;\mathbb{F}) \otimes_{\mathbb{R}} \mathbb{C}$, $V \geq 0$.

b) $$|\det M\{Z\}| \geq c\ |\det M\{iI\}|$$

<u>for all</u> $Z \in V_\varepsilon(n;\mathbb{F})$ <u>and</u> $M \in \Sigma_n$.

<u>Proof.</u> a) Let $\varepsilon \leq 1$, then Proposition 2.4 yields

$$\begin{aligned}|\det(Z+W)| &\geq 2^{-\frac{1}{2}n}(1+n^2\varepsilon^{-2})^{-n}\ |\det(i\varepsilon I+W)|\\ &\geq 2^{-n}(1+n^2\varepsilon^{-2})^{-2n}\ |\det(iI+W)| .\end{aligned}$$

b) According to Corollary II.1.5 it suffices to consider $M = \begin{pmatrix} A & B\\ C & D\end{pmatrix} \in \Sigma_n$, where $C \in GL(n;\mathbb{F})$. In this case the assertion is equivalent to

$$|\det(Z+S)| \geq c\ |\det(iI+S)| ,$$

where $S = C^{-1}D \in \mathrm{Sym}(n;\mathbb{F})$. Part a) completes the proof. □

Next consider $M \in \Sigma_{n,j}$.

<u>Lemma 2.6.</u> a) <u>Given</u> $M \in \Sigma_{n,j}$ <u>of the form</u> (2) <u>one has</u>

$$\det(\operatorname{Im} M\langle Z\rangle_1)\ |\det M\{Z\}|^2 = \det Y_1\ |\det \hat{D}_4|^{2/r} .$$

b) <u>Let</u> $\varepsilon > 0$, <u>then there exists a</u> $c = c(n,\varepsilon) > 0$ <u>satisfying</u>

$$\det(\operatorname{Im} M\langle Z\rangle_1)\ |\det M\{Z\}|^2$$
$$\geq c \det(\operatorname{Im} M\langle iI\rangle_1)\ |\det M\{iI\}|^2$$

<u>for all</u> $Z \in V_\varepsilon(n;\mathbb{F})$ <u>and</u> $M \in \Sigma_n$.

<u>Proof.</u> a) If $M \in \Sigma_{n,j}$ then $\operatorname{Im} M\langle Z\rangle_1 = \operatorname{Im} M_1\langle Z_1\rangle$ and $|\det M\{Z\}|^r = |\det M_1\{Z_1\}|^r\ |\det \hat{D}_4|$ hold in view of Proposition 2.1. Now apply Theorem II.1.7.

b) Observing $\hat{Z} \in V_\varepsilon(rn;\mathbb{R})$, $\hat{M} \in \Sigma_{rn,rj}$ we may assume that $\mathbb{F} = \mathbb{R}$. In the case $j = n$ one has

$$\det(\operatorname{Im} M\langle Z\rangle)\ |\det M\{Z\}|^2 = \det Y \geq \varepsilon^n .$$

If $j = 0$ then

$$|\det M\{Z\}| \geq c|\det M\{iI\}|$$

follows from Lemma 2.5. Let $0 < j < n$. We may multiply M by a matrix in $\Sigma_{n,j}$ from the left. By means of a theorem of the symplectic geometry (cf. SIEGEL [55], Lemma 2) we can transform $M\langle iI\rangle_1$ onto $iI^{(j)}$ and $M\langle Z\rangle_1$ onto iT , where $T \geq I$ is a real diagonal matrix. Lemma 2.5 completes the proof.

□

In order to apply CAUCHY's integral test we need

<u>Lemma 2.7.</u> <u>Given a compact subset</u> C <u>in</u> $H(n;\mathbb{F})$ <u>there exists a constant</u> $c = c(C)$ <u>such that</u>

$$\operatorname{Im} M\langle Z\rangle \leq c \operatorname{Im} M\langle W\rangle$$

<u>holds for all</u> $Z, W \in C$ <u>and</u> $M \in \Sigma_n$.

<u>Proof.</u> There exists $\varepsilon > 0$ satisfying $\varepsilon I \leq Y \leq \varepsilon^{-1} I$ and $X^2 \leq \varepsilon^{-2} I$ for all $Z = X + iY \in C$. In view of Corollary II.1.5 it suffices to show that

$$c^{-1}(\operatorname{Im} M\langle iI\rangle)^{-1} \leq (\operatorname{Im} M\langle Z\rangle)^{-1} \leq c\ (\operatorname{Im} M\langle iI\rangle)^{-1}$$

holds for all $Z \in C$ and $M = \begin{pmatrix} A & B \\ C & D \end{pmatrix} \in \Sigma_n$, where $C \in GL(n;\mathbb{F})$. Putting $S := C^{-1}D \in \operatorname{Sym}(n;\mathbb{F})$ the assertion is equivalent to

$$c^{-1}(I+S^2) \leq Y + Y^{-1}[X+S] \leq c(I+S^2) \ .$$

But $Z \in C$ implies that

$$Y + Y^{-1}[X+S] \leq \varepsilon^{-1}(I + (X+S)^2) \leq 2\varepsilon^{-3}(I+S^2) \ .$$

Replacing S by $S-X$ yields

$$\varepsilon(I+S^2) \leq Y + Y^{-1}[S] \leq 2\varepsilon^{-3}(I + (X-S)^2) \leq 2\varepsilon^{-4}(Y + Y^{-1}[X-S]) \ ,$$

hence the assertion is valid.

□

The notation "$M : \Gamma_{n,j} \setminus \Gamma_n$" indicates that M runs through a set of representatives of the right cosets of Γ_n modulo $\Gamma_{n,j}$, i.e. through a set $\{M_l \ ; \ l \in I\}$ such that

$$\Gamma_n = \bigcup_{l \in I} \Gamma_{n,j} M_l$$

holds, where the cosets are mutually disjoint.

The essential assertion on the behavior of convergence is contained in

Theorem 2.8. *Given* $0 \leq j < n$ *and* $k > 2 + r(n+j-1)$ *then the series*

$$g^k_{n,j}(Z) := \sum_{M:\Gamma_{n,j}\setminus\Gamma_n} (\det \operatorname{Im} M\langle Z\rangle_1)^{-\frac{1}{2}k} \, |\det M\{Z\}|^{-k} \ , \quad Z \in H(n;\mathbb{F}) \ ,$$

converges uniformly in any vertical strip $V_\varepsilon(n;\mathbb{F})$, $\varepsilon > 0$.

Here and in the following, of course, one has to read

$$g^k_{n,o}(Z) = \sum_{M:\Gamma_{n,o}\setminus\Gamma_n} |\det M\{Z\}|^{-k} \ .$$

Proof. The definition does not depend on the choice of the representatives in view of Lemma 2.6 a) and because of part b) it suffices to consider $Z = iI$. Then one has

$$g^k_{n,j}(iI) = \sum_{M:\Gamma_{n,j}\setminus\Gamma_n} (\det \operatorname{Im} M\langle iI\rangle_1)^{-\frac{1}{2}k} \, (\det \operatorname{Im} M\langle iI\rangle)^{\frac{1}{2}k} \ .$$

Put $M\langle Z\rangle = x(Z) + iy(Z)$ and let C denote a compact subset of $F(n;\mathbb{F})$ possessing positive symplectic volume. If we apply Lemma 2.7 twice we can find $c_1 > 0$ such that

$$(\det y_1(iI))^{-\frac{1}{2}k} (\det y(iI))^{\frac{1}{2}k} \leq c_1 (\det y_1(Z))^{-\frac{1}{2}k} (\det y(Z))^{\frac{1}{2}k}$$

holds for all $Z \in C$. Let dv_n denote the symplectic volume element. Putting $c_2 = c_1 (\int_C dv_n)^{-1}$ yields

$$\begin{aligned} &(\det y_1(iI))^{-\frac{1}{2}k} (\det y(iI))^{\frac{1}{2}k} \\ &\leq c_2 \int_C (\det y_1(Z))^{-\frac{1}{2}k} (\det y(Z))^{\frac{1}{2}k} dv_n \\ &= c_2 \int_{M\langle C\rangle} (\det Y_1)^{-\frac{1}{2}k} (\det Y)^{\frac{1}{2}k} dv_n . \end{aligned}$$

Since C is compact, there exists $t := \max\{\det Y;\ Z = X + iY \in C\}$. In view of $C \subset F(n;\mathbb{F})$ one has $\det Y \leq t$ for all $Z = X + iY$ belonging to $\bigcup_{M:\Gamma_{n,j}\setminus\Gamma_n} M\langle C\rangle$. Let l denote the number of matrices $M \in \Gamma(n;\mathbb{F})$ satisfying $M\langle F(n;\mathbb{F})\rangle \cap F(n;\mathbb{F}) \neq \emptyset$. According to Proposition 2.2 it suffices to demonstrate the convergence of the integral

$$\int_{F_{n,j},\det Y \leq t} (\det Y_1)^{-\frac{1}{2}k} (\det Y)^{\frac{1}{2}k} dv_n ,$$

since this domain is covered at most l-times. If $j = 0$ the integral

$$\int_{R(n;\mathbb{F}),\det Y \leq t} (\det Y)^{\frac{1}{2}k-2h/n} dY$$

converges according to Corollary I.5.10, whenever $k > \frac{2h}{n} = 2 + r(n-1)$. Let $0 < j < n$. Given $Y = \begin{pmatrix} Y_1 & 0 \\ 0 & Y_2 \end{pmatrix} \begin{bmatrix} I & Y_3 \\ 0 & I \end{bmatrix}$ one has

$$\begin{aligned} dv_n &= (\det Y)^{-2h/n} dX\, dY \\ &= (\det Y_1)^{-2-r(j-1)} (\det Y_2)^{-2-r(n-1)} dX\, dY_1\, dY_2\, dY_3 . \end{aligned}$$

In view of Proposition II.3.9 the volume

$$\int_{F(j;\mathbb{F})} dv_j$$

is finite. Proposition II.3.1 yields $\det(\mathrm{Im}\, Z_1) \geq c$ for $Z_1 \in F(j;\mathbb{F})$. After estimating the integral over all compact components it suffices to consider the integral

$$\int_{\substack{R(n-j;\mathbb{F}) \\ \det Y_2 \leq t}} (\det Y_2)^{\frac{1}{2}k-r(n-1)-2} dY_2 .$$

According to Corollary I.5.10 this integral converges, whenever $k > 2 + r(n+j-1)$. □

Consider the formal extension

$$[\Gamma(0;\mathbb{F}),k] = [\Gamma(0;\mathbb{F}),k]_0 := \mathbb{C} .$$

Given $n \geq 1$, $0 \leq j \leq n$ and a cusp form $f \in [\Gamma(j;\mathbb{F}),k]_0$ we define

$$E^k_{n,j}(Z,f) := \sum_{M:\Gamma_{n,j}\setminus\Gamma_n} f(M\langle Z\rangle_1)\ (\det M\{Z\})^{-k} , \quad Z \in H(n;\mathbb{F}) .$$

$E^k_{n,j}(Z,f)$ is called EISENSTEIN-series in Z with respect to f .

Obviously,

$$E^k_{n,n}(\cdot,f) = f$$

holds. In general, we formulate

Theorem 2.9. *Let $0 \leq j < n$, $k > 2 + r(n+j-1)$, $k \equiv 0 \bmod s(\mathbb{F})$. Given a cusp form $f \in [\Gamma(j;\mathbb{F}),k]_0$ the EISENSTEIN-series $E^k_{n,j}(\cdot,f)$ converges absolutely and uniformly in every vertical strip $V_\varepsilon(n;\mathbb{F})$, $\varepsilon > 0$, and represents a modular form on $H(n;\mathbb{F})$ of weight k .*

Proof. Given $K \in \Gamma_{n,j}$ Proposition 2.1 yields $KM\langle Z\rangle_1 = K_1\langle M\langle Z\rangle_1\rangle$ and $(\det KM\{Z\})^{-k} = (\det K_1\{M\langle Z\rangle_1\})^{-k}\ (\det M\{Z\})^{-k}$ because of $k \equiv 0 \bmod s(\mathbb{F})$. We obtain $f(KM\langle Z\rangle_1) = (\det K_1\{M\langle Z\rangle_1\})^k f(M\langle Z\rangle_1)$ from $f \in [\Gamma(j;\mathbb{F}),k]_0$. Hence the definition does not depend on the choice of the representatives.

According to Lemma III.2.4 the function $(\det Y_1)^{\frac{1}{2}k}\ |f(Z_1)|$ is invariant under $\Gamma(j;\mathbb{F})$ and bounded, because f is a cusp form. Thus it suffices to prove the uniform convergence of the series

$$\sum_{M:\Gamma_{n,j}\setminus\Gamma_n} (\det \operatorname{Im} M\langle Z\rangle_1)^{-\frac{1}{2}k}\ |\det M\{Z\}|^k$$

in $V_\varepsilon(n;\mathbb{F})$. This was done in Theorem 2.8. Hence $E^k_{n,j}(\cdot,f)$ is also a holomorphic function and the condition of boundedness in the case $n = 1$ is fulfilled, too.

Given $K \in \Gamma_n$ Theorem II.1.7 yields

$$E^k_{n,j}(K\langle Z\rangle,f) = \sum_{M:\Gamma_{n,j}\setminus\Gamma_n} f(MK\langle Z\rangle_1)\ (\det M\{K\langle Z\rangle\})^{-k}$$

$$= (\det K\{Z\})^k \sum_{M:\Gamma_{n,j}\setminus\Gamma_n} f(MK\langle Z\rangle_1)\ (\det MK\{Z\})^{-k}$$

$$= (\det K\{Z\})^k\ E^k_{n,j}(Z,f) \ ,$$

since MK also runs through a set of representatives and the series may be rearranged. Thus we have

$$E^k_{n,j}(\cdot,f) \in [\Gamma(n;\mathbb{F}),k] \ .$$

□

Next we want to describe the effect of SIEGEL's ϕ-operator. But first a straightforward computation leads to

Proposition 2.10. Let $n \geq 2$ and $0 \leq j < n$. Then the map

$$\Gamma_{n-1,j}\diagdown^{\Gamma_{n-1}} \longrightarrow \Gamma_{n,j} \cap \Gamma_{n,n-1}\diagdown^{\Gamma_{n,n-1}} \ ,$$

$$\Gamma_{n-1,j}\, M \longmapsto (\Gamma_{n,j} \cap \Gamma_{n,n-1})(M \times I^{(2)}) \ ,$$

is bijective.

Using the formal definition $E^k_{o,o}(\cdot,f) = f$ one has

Theorem 2.11. Let $0 \leq j < n$, $k > 2 + r(n+j-1)$ and $k \equiv 0 \bmod s(\mathbb{F})$. Every cusp form $f \in [\Gamma(j;\mathbb{F}),k]_o$ fulfills

$$E^k_{n,j}(\cdot,f)\,|\,\phi = E^k_{n-1,j}(\cdot,f) \ ,$$

especially,

$$E^k_{n,j}(\cdot,f)\,|\,\phi^{n-j} = f \ .$$

$E^k_{n,j}(\cdot,f)$ vanishes identically if and only if $f \equiv 0$.

Proof. In view of the uniform convergence we may distribute the limit through the infinite series. Let $n = 1$ and $M = \begin{pmatrix} a & b \\ c & d \end{pmatrix} \in \Gamma(1;\mathbb{R})$, then

$$\lim_{\lambda\to\infty} (M\{i\lambda\})^{-1} = \lim_{\lambda\to\infty} (ic\lambda+d)^{-1} = \begin{cases} d^{-1} & \text{for } c = 0 \\ 0 & \text{for } c \neq 0 \end{cases} .$$

It follows that $E^k_{1,o}(\cdot,f)\,|\,\phi = f$.

Let $n \geq 2$ and $Z_\lambda := [Z, i\lambda]$, where $Z \in H(n-1;\mathbb{F})$. Given $M \in \Gamma_{n,n-1}$ the term

$$f(M\langle Z_\lambda\rangle_1)\ (\det M\{Z_\lambda\})^{-k}$$

does not depend on λ. Proposition 2.10 yields

$$\sum_{M:\Gamma_{n,j}\cap\Gamma_{n,n-1}\setminus\Gamma_{n,n-1}} f(M\langle Z_\lambda\rangle_1)(\det M\{Z_\lambda\})^{-k} = E^k_{n-1,j}(Z,f) .$$

Thus it suffices to show that

$$\lim_{\lambda\to\infty} \det(\operatorname{Im} M\langle Z_\lambda\rangle_1)\ (\det M\{Z_\lambda\})^2 = \infty$$

holds, whenever $M \in \Gamma_n$ and $\Gamma_{n,j}M \cap \Gamma_{n,n-1} = \emptyset$, if we use the same estimations for f as in the proof of Theorem 2.9. Put

$$M = \begin{pmatrix} * & * \\ C & D\end{pmatrix} , \quad C = \begin{pmatrix} C_1 & C_2 \\ C_3 & C_4\end{pmatrix} , \quad D = \begin{pmatrix} D_1 & D_2 \\ D_3 & D_4\end{pmatrix} ,$$

where $C_1 = C_1^{(j,n-1)}$, $D_1 = D_1^{(j,n-1)}$. If $Z_\lambda = X_\lambda + iY_\lambda$ Theorem II.1.7 yields

$$(\operatorname{Im} M\langle Z_\lambda\rangle)^{-1} = Y_\lambda[\bar{C}'] + Y_\lambda^{-1}[X_\lambda\bar{C}' + \bar{D}'] ,$$

$$\det(\operatorname{Im} M\langle Z_\lambda\rangle_1)\ |\det M\{Z_\lambda\}|^2 = \det Y_\lambda \det V ,$$

where $(\operatorname{Im} M\langle Z_\lambda\rangle)^{-1} = \begin{pmatrix} * & * \\ * & V\end{pmatrix}$, $V \in \operatorname{Pos}(n-j;\mathbb{F})$. Putting $S = Y[\bar{C}_3'] + Y^{-1}[X\bar{C}_3' + \bar{D}_3']$ leads to

$$p(\lambda) := \det Y_\lambda \det V = \lambda \det Y \det(S + \lambda C_4\bar{C}_4' + \lambda^{-1}D_4\bar{D}_4') .$$

Since C_4 and D_4 are column vectors, we conclude from Theorem I.3.4 that $p(\lambda)$ is a real polynomial. Thus we have to show that p is not constant. If p were constant $S + \lambda C_4\bar{C}_4' + \lambda^{-1}D_4\bar{D}_4' \geq S$ would imply

$$p(\lambda) \geq \lambda \det Y \det S .$$

We obtain $\det S = 0$ from $\det Y > 0$. Because of $S = Y[\bar{C}_3'] + Y^{-1}[X\bar{C}_3' + \bar{D}_3']$ the rank of $(C_3 D_3)$ cannot be maximal. Multiplying $(C_3 D_3)$ by a unimodular matrix from the left, i.e. choosing a suitable representative of $\Gamma_{n,j}M$, we may suppose that the last row of $(C_3 D_3)$ vanishes. Let c resp. c denote the last element in the column C_4 resp. D_4. Lemma II.1.1 implies that $c\bar{D}_4' = d\bar{C}_4'$ and $c\bar{d} = d\bar{c}$ hold. If $c \neq 0$ it would follow that

$$\det(S + \lambda C_4\bar{C}_4' + \lambda^{-1}D_4\bar{D}_4') = \det(S + (\lambda + \lambda^{-1}\left|\frac{d}{c}\right|^2)C_4\bar{C}_4') .$$

Hence $p(\lambda)$ would have the zero $\lambda = i\left|\frac{d}{c}\right|$ and would not be constant. On the other hand $c = 0$ yields $M \in \Gamma_{n,n-1}$ and therefore a contradiction.

□

The theorem says that cusp forms can be "lifted" by means of EISENSTEIN-series, whenever the weight is sufficiently large in proportion to the degree.

Theorem 2.12. Let $k > 2 + 2r(n-1)$, $k \equiv 0 \bmod s(\mathbb{F})$. Then SIEGEL's ϕ-operator

$$\phi : [\Gamma(n;\mathbb{F}),k] \longrightarrow [\Gamma(n-1;\mathbb{F}),k]$$

becomes surjective.

Proof. By induction on j we show that the map $\phi^{n-j} : [\Gamma(n;\mathbb{F}),k] \longrightarrow [\Gamma(j;\mathbb{F}),k]$ is surjective. In the case $j = 0$ then $f \in [\Gamma(0;\mathbb{F}),k]$ is a constant, the EISENSTEIN-series $E^k_{n,o}(\cdot,f)$ converges and satisfies $E^k_{n,o}(\cdot,f)|\phi^n = f$ in view of Theorem 2.11.

Let $f \in [\Gamma(j+1;\mathbb{F}),k]$. By induction hypothesis there exists $g \in [\Gamma(n;\mathbb{F}),k]$ such that $f|\phi = g|\phi^{n-j}$. Hence $f_o := g|\phi^{n-j-1} - f$ is a cusp form and the preceding theorem yields

$$(g - E^k_{n,j+1}(\cdot,f_o))|\phi^{n-j-1} = g|\phi^{n-j-1} - f_o = f .$$

□

Thus it also follows that $\dim[\Gamma(n;\mathbb{F}),k] \geq \dim[\Gamma(1;\mathbb{F}),k]$, whenever $k > 2 + 2r(n-1)$, $k \equiv 0 \bmod s(\mathbb{F})$. Using [52], p.88, one has

Corollary 2.13. Let $k > 2 + 2r(n-1)$, $k \equiv 0 \bmod s(\mathbb{F})$, then

$$\dim [\Gamma(n;\mathbb{F}),k] = \dim[\Gamma(n;\mathbb{F}),k]_o + \dim[\Gamma(n-1;\mathbb{F}),k] \geq \begin{cases} \left[\frac{k}{12}\right] & \text{for } k \equiv 2 \bmod 12 \\ \left[\frac{k}{12}\right] + 1 & \text{else} \end{cases} .$$

Putting $j = 0$, $f = 1$ we have the classical EISENSTEIN-series

$$E^k_n(Z) := E^k_{n,o}(Z,1) \ , \ Z \in H(n;\mathbb{F}) \ ,$$

introduced by SIEGEL [53]. Theorem 2.9 says that E^k_n converges absolutely, whenever $k > 2 + r(n-1)$. BRAUN [6],[7] proved that one cannot improve this bound, if one considers SIEGEL's or Hermitian modular forms. Analogous arguments lead to

Theorem 2.14. The EISENSTEIN-series E_n^k does not converge absolutely for $k = 2 + r(n-1)$.

Proof. We show that E_n^k does not converge absolutely in the point iI. Therefore we let M run through a part of special set of representatives of the right cosets. Consider the matrices

$$C = \begin{pmatrix} c & 0 \\ \mathfrak{c} & I \end{pmatrix}, \quad D = \begin{pmatrix} d & -d\bar{\mathfrak{c}}' \\ 0 & 0 \end{pmatrix}, \quad \mathfrak{c} = (c_2,\dots,c_n)',$$

where $c \in \mathbb{N}$, $c_1 \in \mathcal{O} \bmod c$ and $d \in \mathbb{N}$, $1 \leq d \leq c$, relatively prime to c. Choose $a,b \in \mathbb{Z}$ satisfying $ad - bc = 1$ and define

$$A = \begin{pmatrix} a & 0 \\ 0 & 0 \end{pmatrix}, \quad B = \begin{pmatrix} b & -b\bar{\mathfrak{c}}' \\ 0 & -I \end{pmatrix}.$$

Then $M = \begin{pmatrix} A & B \\ C & D \end{pmatrix}$ belongs to $\Gamma(n;\mathbb{F})$ and the right cosets $\Gamma_{n,0} M \begin{pmatrix} I & S \\ 0 & I \end{pmatrix}$, where M runs through the set quoted above and S through $\mathrm{Sym}(n;\mathcal{O})$, are mutually disjoint.

Now consider the corresponding partial series

$$g_n^k = \sum_{M,S} \left| \det M \begin{pmatrix} I & S \\ 0 & I \end{pmatrix} \{iI\} \right|^{-k}.$$

One has $M \begin{pmatrix} I & S \\ 0 & I \end{pmatrix} \{iI\} = C(iI + C^{-1}D + S)$. The series

$$\sum_{S \in \mathrm{Sym}(n;\mathcal{O})} |\det(iI + C^{-1}D + S)|^{-k}$$

converges according to Corollary 1.5, whenever $k > 1 + r(n-1)$. Thus the series $\sum_M c^{-k}$, $k = 2 + r(n-1)$, would have to converge. Since the set $x \in \mathcal{O} \bmod c$ consists of c^r elements one calculates that

$$\sum_M c^{-k} = \sum_{c=1}^{\infty} \varphi(c)\, c^{-2},$$

where φ denotes EULER's totient function. But the last series diverges. Hence E_n^k, $k = 2 + r(n-1)$, cannot converge absolutely.

□

What about the case $n = 1$? If $k > 2$ is even (resp. $k \equiv 0 \bmod 4$ for $\mathbb{F} = \mathbb{C}$) one has

$$E_1^k(z) = \tfrac{1}{2} \sum_{(c,d)} (cz + d)^{-k}, \quad z \in H(1;\mathbb{F}),$$

where (c,d) runs through the set of relatively prime pairs in $\mathbb{Z} \times \mathbb{Z}$.

§3 POINCARE´-series

In this paragraph we study generalizations of POINCARE´-series introduced in the case of SIEGEL's modular group by KLINGEN [31]. According to [33] we can reduce the behavior of convergence to that of EISENSTEIN-series. A suitable FOURIER-expansion establishes a connection to another type of POINCARE´-series investigated by MAASS [43].

Lemma II.1.1 and Theorem II.1.7 immediately lead to

<u>Proposition 3.1.</u> <u>Let</u> $M = \begin{pmatrix} A & B \\ C & D \end{pmatrix} \in \Sigma_n$, $Z,W \in H(n;\mathbb{F})$ <u>and put</u>

$$M\{Z,W\} = AZ + B + WCZ + WD = (M\langle Z\rangle + W)\, M\{Z\} ,$$

$$M^* := \begin{pmatrix} A & -B \\ -C & D \end{pmatrix} = M\begin{bmatrix} I & 0 \\ 0 & -I \end{bmatrix} \in \Sigma_n .$$

<u>Given</u> $M, M_o \in \Sigma_n$ <u>the following holds:</u>

a) $\overline{M\{Z,W\}}' = M^{*-1}\{W,Z\}$,

b) $MM_o\{Z,W\} = M\{M_o\langle Z\rangle ,W\}\, M_o\{Z\}$,

c) $M\langle -\tilde{Z}\rangle = -\widetilde{M^*\langle Z\rangle}$,

d) $(\det M\{-\tilde{Z}\})^2 = \overline{\det M^*\{Z\}^2}$.

We need certain subgroups of the modular group. Put

$$A_{n,n} := \left\{\begin{pmatrix} I & S \\ O & I \end{pmatrix} ;\ S \in \mathrm{Sym}(n;\mathcal{O})\right\} .$$

If $0 < j \leq n$, consider the homomorphism $\chi : \Gamma_{n,j} \longrightarrow \Gamma_j$ introduced in Proposition 2.1 and define

$$A_{n,j} := \chi^{-1}(A_{j,j}) ,$$

$$B_{n,j} := \text{kernel } \chi .$$

In the case $j = 0$ put $A_{n,o} = B_{n,o} = \Gamma_{n,o}$. Now

$$\Gamma_{n,o} = \left\{ \begin{pmatrix} U & S\bar{U}'^{-1} \\ O & \bar{U}'^{-1} \end{pmatrix} ;\ U \in GL(n;\mathcal{O}) ,\ S \in \mathrm{Sym}(n;\mathcal{O}) \right\}$$

and the description of $\Gamma_{n,j}$ in Proposition 2.1 lead to

Proposition 3.2. Given $0 \leq j \leq n$ one has

$$A_{n,j} = \left\{ \begin{pmatrix} U & S\bar{U}'^{-1} \\ 0 & \bar{U}'^{-1} \end{pmatrix} ; \; U = \begin{pmatrix} I^{(j)} & 0 \\ * & * \end{pmatrix} \in GL(n;\mathcal{O}) \;, \; S \in Sym(n;\mathcal{O}) \right\},$$

$$B_{n,j} = \left\{ \begin{pmatrix} U & S\bar{U}'^{-1} \\ 0 & \bar{U}'^{-1} \end{pmatrix} \in A_{n,j} \; ; \; S_1 = 0 \right\}.$$

According to KLINGEN [33] we prove

Lemma 3.3. Let $k > 2 + 2r(n-1)$ and $\varepsilon > 0$, then there exists a constant $c = c(n,k,\varepsilon) > 0$ satisfying

$$\sum_{M \in \Gamma_n} |\det M\{Z,W\}|^{-k} \leq c \, (\det Y)^{-\frac{1}{2}k}$$

for all $Z \in H(n;\mathbb{F})$ and $W \in V_\varepsilon(n;\mathbb{F})$.

Proof. In view of Lemma 2.5 it suffices to consider $W = iI$. Choosing a suitable compact subset $C \subset F(n;\mathbb{F})$ with positive symplectic volume one can find a constant $c_1 = c_1(n) > 0$ such that

$$|\det(Z+W)| \, (\det V)^{-\frac{1}{2}} \leq c_1 |\det(Z+iI)|$$

holds for all $W = U + iV \in C$ and all $Z \in H(n;\mathbb{F})$ according to Proposition 2.4. Thus there exists $c_2 = c_2(n) > 0$ satisfying

$$(1) \qquad |\det(Z+iI)|^{-k} \leq c_2^{-k} \int_{F(n;\mathbb{F})} |\det(Z+W)|^{-k} \, (\det V)^{\frac{1}{2}k} \, dv(W) \quad .$$

Proposition 3.1 yields

$$\int_F |\det M\{Z,W\}|^{-k} \, (\det V)^{\frac{1}{2}k} \, dv(W)$$

$$= \int_F |\det M^{*-1}\{W,Z\}|^{-k} \, (\det V)^{\frac{1}{2}k} \, dv(W)$$

$$= \int_F |\det(M^{*-1}\langle W\rangle + Z)|^{-k} \, (\det \operatorname{Im} M^{*-1}\langle W\rangle)^{\frac{1}{2}k} \, dv(W)$$

$$= \int_{M^{*-1}\langle F\rangle} |\det(W+Z)|^{-k} \, (\det V)^{\frac{1}{2}k} \, dv(W) \; ,$$

where $F = F(n;\mathbb{F})$. Put $\gamma = \operatorname{ord} E(\mathbb{F})$, if $n = 1$, and $\gamma = s(\mathbb{F})$, if $n > 1$. Since $k > 2 + 2r(n-1)$ Corollary 1.3 and (1) lead to

$$\sum_{M\in\Gamma_n} |\det M\{Z,iI\}|^{-k}$$

$$\leq c_2^{-k} \sum_{M\in\Gamma_n} \int_F |\det M\{Z,W\}|^{-k} (\det V)^{\frac{1}{2}k} dv(W)$$

$$= c_2^{-k} \sum_{M\in\Gamma_n} \int_{M^{*-1}<F>} |\det(W+Z)|^{-k} (\det V)^{\frac{1}{2}k} dv(W)$$

$$= c_2^{-k} \gamma \int_{H(n;\mathbb{F})} |\det(W+Z)|^{-k} (\det V)^{\frac{1}{2}k} dv(W)$$

$$= \alpha c_2^{-k} \gamma (\det Y)^{-\frac{1}{2}k} .$$

□

Let $0 \leq j \leq n$ and let $\varphi : H(j;\mathbb{F}) \longrightarrow \mathbb{C}$ be holomorphic and bounded. Given $k \in \mathbb{N}$, where $k \equiv 0 \bmod 2$ for $\mathbb{F} = \mathbb{H}$, $Z \in H(n;\mathbb{F})$ and $W \in H(j;\mathbb{F})$ we define

$$P_{n,j}^k(Z,W,\varphi) := \sum_{M:\mathcal{B}_{n,j}\setminus\Gamma_n} \varphi(M<Z>_1) \det(M<Z>_1 + W)^{-k} (\det M\{Z\})^{-k}$$

and call $P_{n,j}^k(Z,W,\varphi)$ the POINCARE´-<u>series with respect to</u> φ <u>and developmental center</u> W.

In the case $j = 0$ we look upon φ as a constant and have

$$P_{n,o}^k(Z,W,\varphi) = E_{n,o}^k(Z,\varphi) = \varphi\, E_n^k(Z) .$$

<u>Theorem 3.4.</u> <u>Let</u> $\varphi : H(n;\mathbb{F}) \longrightarrow \mathbb{C}$ <u>be holomorphic and bounded</u>, $k > 2 + 2r(n-1)$ <u>and</u> $k \equiv 0 \bmod 2$, <u>if</u> $\mathbb{F} = \mathbb{H}$. <u>Then the</u> POINCARE´-<u>series</u>

$$P_{n,n}^k(Z,W,\varphi) := \sum_{M\in\Gamma_n} \varphi(M<Z>) (\det M\{Z,W\})^{-k}$$

<u>converges absolutely and uniformly for</u> $Z,W \in V_\varepsilon(n;\mathbb{F})$, $\varepsilon > 0$, <u>and represents a cusp form on</u> $H(n;\mathbb{F})$ <u>of weight</u> k.

<u>Proof.</u> Since φ is bounded, the series of absolute values is majorized by

$$\sum_{M\in\Gamma_n} |M\{iI,iI\}|^{-k}$$

for $Z,W \in V_\varepsilon(n;\mathbb{F})$ in view of Lemma 2.5 and Proposition 3.1. The last series converges according to the preceding lemma. Given $M_0 \in \Gamma_n$ Proposition 3.1 yields

$$P^k_{n,n}(M_0\langle Z\rangle,W,\varphi) = (\det M_0\{Z\})^k \sum_{M\in\Gamma_n} \varphi(MM_0\langle Z\rangle)(\det MM_0\{Z,W\})^{-k}$$

$$= (\det M_0\{Z\})^k P^k_{n,n}(Z,W,\varphi) ,$$

since we may rearrange the series in view of its absolute convergence. The properties of boundedness in the case $n = 1$ and of being a cusp form immediately follow from Lemma 3.3.

□

In contrast to EISENSTEIN- and theta-series it is a difficult problem to decide whether POINCARE´-series identically vanish. A connection between EISENSTEIN- and POINCARE´-series is established by

Theorem 3.5. Let $0 < j < n$, $k > 2 + r(n+j-1)$, $k \equiv 0 \bmod s(\mathbb{F})$, $\varepsilon > 0$. If $\varphi : H(j;\mathbb{F}) \longrightarrow \mathbb{C}$ is holomorphic and bounded, the POINCARE´-series

$$P^k_{n,j}(Z,W,\varphi) := \sum_{M:\mathcal{B}_{n,j}\setminus\Gamma_n} \varphi(M\langle Z\rangle_1) \det(M\langle Z\rangle_1+W)^{-k} (\det M\{Z\})^{-k}$$

converges absolutely and uniformly for $Z \in V_\varepsilon(n;\mathbb{F})$, $W \in V_\varepsilon(j;\mathbb{F})$ and represents a modular form on $H(n;\mathbb{F})$ of weight k . One has

$$E^k_{n,j}(\cdot,P^k_{j,j}(\cdot,W,\varphi)) = P^k_{n,j}(\cdot,W,\varphi)$$

and

$$P^k_{n,j}(\cdot,W,\varphi)\,|\,\phi = P^k_{n-1,j}(\cdot,W,\varphi) .$$

Proof. Proposition 3.2, Proposition 2.1 and $\mathcal{B}_{n,j}$ = kernel χ yield that the definition of $P^k_{n,j}(Z,W,\varphi)$ does not depend on the choice of the representatives, whenever $k \equiv 0 \bmod s(\mathbb{F})$. In view of Theorem 3.4 and Theorem 2.9

$$f(Z) := E^k_{n,j}(Z,P^k_{j,j}(\cdot,W,\varphi)) , \quad Z \in H(n;\mathbb{F}) ,$$

becomes a modular form on $H(n;\mathbb{F})$ of weight k . The boundedness of φ , Lemma 3.3 and Theorem 2.8 imply that the series

$$f(Z) = \sum_{\substack{M:\Gamma_{n,j}\setminus\Gamma_n \\ K_1\in\Gamma_j}} \varphi(K_1\langle M\langle Z\rangle_1\rangle)(\det K_1\{M\langle Z\rangle_1,W\})^{-k}(\det M\{Z\})^{-k}$$

converges absolutely and uniformly for $Z \in V_\varepsilon(n;\mathbb{F})$ and $W \in V_\varepsilon(j;\mathbb{F})$. Proposition 3.1 and Proposition 2.1 yield

$$f(Z) = \sum_{\substack{M:\Gamma_{n,j}\setminus\Gamma_n \\ K_1\in\Gamma_j, K=K_1\times I}} \varphi(KM\langle Z\rangle_1)\, \det(KM\langle Z\rangle_1+W)^{-k}\, (\det KM\{Z\})^{-k} .$$

But KM runs through a set of representatives of $B_{n,j}\setminus\Gamma_n$, hence

$$f(Z) = P^k_{n,j}(Z,W,\varphi) .$$

Theorem 2.11 completes the proof.

□

We pay special attention to the case $\varphi \equiv 1$ and put

$$P^k_{n,j}(Z,W) := P^k_{n,j}(Z,W,1) .$$

<u>Theorem 3.6.</u> <u>Let</u> $1 \leq j \leq n$, $k > \min\{2+r(n+j-1), 2+2r(n-1)\}$, $k \equiv 0 \bmod s(\mathbb{F})$. <u>Given</u> $W \in H(j;\mathbb{F})$ <u>then</u> $P^k_{n,j}(\cdot,W)$ <u>is a modular form on</u> $H(n;\mathbb{F})$ <u>of weight</u> k. <u>Given</u> $Z \in H(n;\mathbb{F})$ <u>then</u> $P^k_{n,j}(Z,\cdot)$ <u>becomes a cusp form on</u> $H(j;\mathbb{F})$ <u>of weight</u> k. <u>One has</u>

$$P^k_{n,n}(Z,W) = P^k_{n,n}(W,Z) ,$$

$$P^k_{n,j}(-\tilde{Z},-\tilde{W}) = \overline{P^k_{n,j}(Z,W)} .$$

<u>If</u> $k > 2+r(n-1)$, $k \equiv 0 \bmod s(\mathbb{F})$, <u>then</u> $E^k_n(-\tilde{Z}) = \overline{E^k_n(Z)}$ <u>holds. Especially, the</u> FOURIER-<u>coefficients of the modular forms</u> $P^k_{n,j}(\cdot,iV)$, $V \in Pos(j;\mathbb{F})$, $P^k_{n,j}(iY,\cdot)$, $Y \in Pos(n;\mathbb{F})$ <u>and</u> E^k_n <u>turn out to be real numbers</u>.

<u>Proof.</u> In view of Theorem 3.4 and Theorem 3.5 the POINCARE'-series $P^k_{n,j}(\cdot,W)$ is a modular form of weight k. Since k is even one has $(\det M\{Z\})^{-k} = (\det \overline{M\{Z\}}')^{-k}$. Given $M \in \Gamma_n$, $K_1 \in \Gamma_j$, where $K = K_1 \times I$, we obtain

$$\det(M\langle Z\rangle_1 + K_1\langle W\rangle)^{-k}\, (\det M\{Z\})^{-k}$$

$$= (\det K_1\{W\})^k\, \det(K^{*-1}M\langle Z\rangle_1+W)^{-k}\, (\det K^{*-1}M\{Z\})^{-k}$$

from Proposition 3.1. Since $B_{n,j}$ is a normal subgroup of $\Gamma_{n,j}$, together with M also $K^{*-1}M$ runs through a set of representatives of $B_{n,j}\setminus\Gamma_n$. A rearrangement yields

$$P^k_{n,j}(Z,K_1\langle W\rangle) = (\det K_1\{W\})^k\, P^k_{n,j}(Z,W) \ .$$

Because of the uniform convergence in vertical strips we may distribute the limit through the infinite series, if we compute the effect of SIEGEL's ϕ-operator. In view of

$$\lim_{\lambda\to\infty} \det(M\langle Z\rangle_1 + W_\lambda)^{-k} = 0 \quad \text{for} \quad W_\lambda = [W_o, i\lambda] \ ,$$

$P^k_{n,j}(Z,\cdot)$ is a cusp form.

Because of $k \equiv 0 \bmod s(\mathbb{F})$ Proposition 3.1 leads to

$$(\det M\{Z,W\})^{-k} = (\det M^{*-1}\{W,Z\})^{-k} \ ,$$

whenever $Z,W \in H(n;\mathbb{F})$ and $M \in \Gamma_n$. A rearrangement yields

$$P^k_{n,n}(Z,W) = P^k_{n,n}(W,Z) \ .$$

Since k is even, we obtain

$$\det(M\langle -\tilde{Z}\rangle_1 - \tilde{W})^{-k}\,(\det M\{-\tilde{Z}\})^{-k}$$
$$= \overline{(\det(M^*\langle Z\rangle_1 + W))^{-k}(\det M^*\{Z\})^{-k}}$$

from Proposition 3.1. Together with M also M^* runs through a set of representatives of $\mathcal{B}_{n,j}\backslash\Gamma_n$. Hence we have

$$P^k_{n,j}(-\tilde{Z},-\tilde{W}) = \overline{P^k_{n,j}(Z,W)} \quad \text{and} \quad E^k_n(-\tilde{Z}) = \overline{E^k_n(Z)}$$

in the case $j = 0$. Consider $P^k_{n,j}(-\tilde{Z},iV) = \overline{P^k_{n,j}(Z,iV)}$ resp. $P^k_{n,j}(iY,-\tilde{W}) = \overline{P^k_{n,j}(iY,W)}$. Now it follows from the uniqueness of the FOURIER-expansion that the FOURIER-coefficients are real numbers.

□

If we consider the FOURIER-expansion of $P^k_{n,j}(Z,W)$ with respect to W, the FOURIER-coefficients form another type of POINCARE'-series as functions of Z, which also turn out to be modular forms of weight k. These POINCARE'-series were investigated by MAASS [43].

Given $k \in \mathbb{N}$, $1 \leq j \leq n$ and a positive definite $T \in Sym^\tau(j;\mathcal{O})$ by a POINCARE'-<u>series in</u> T we mean

$$Q^k_{n,j}(Z,T) = \sum_{M:\mathcal{A}_{n,j}\backslash\Gamma_n} e^{2\pi i\tau(T,M\langle Z\rangle_1)}\,(\det M\{Z\})^{-k} \ , \quad Z \in H(n;\mathbb{F}).$$

In the case $j = 0$ consider the formal extension

$$Q^k_{n,o}(Z,T) = E^k_n(Z) \ .$$

In order to examine the behavior of convergence we need

Lemma 3.7. Let $\varepsilon > 0$, $k > 0$ and $\mathcal{D} := \bigcup_{M \in A_{n,n}} M\langle\mathcal{F}(n;\mathbb{F})\rangle$. Then there exists a constant $c = c(k,n,\varepsilon) > 0$ such that

$$e^{-2\pi\varepsilon \operatorname{tr}(Y)} \leq c \int_{\mathcal{D}} |\det(Z+W)|^{-k} (\det V)^{\frac{1}{2}k} \, dv(W)$$

holds for all $Z = X + iY \in H(n;\mathbb{F})$.

Proof. It suffices to show that there exists a constant c_1 satisfying

$$|\det(Z+W)|^k \leq c_1 \, e^{2\pi\varepsilon \operatorname{tr}(Y)},$$

whenever W belongs to a compact subset of $\mathcal{F}(n;\mathbb{F})$ having finite symplectic volume and $Z = X + iY \in H(n;\mathbb{F})$, where $X \in \mathcal{C}(n;\mathbb{F})$. We obtain a constant c_2 such that

$$|\det(Z+W)|^k \leq c_2^k \, (1 + \operatorname{tr}(Y))^{nk} \, (\det V)^k$$

holds for all these Z,W in view of Proposition 2.4. Since $(1 + \operatorname{tr}(Y))^{nk} \, e^{-2\pi\varepsilon \operatorname{tr}(Y)}$ remains bounded for $Y > 0$, the assertion follows.

□

Now we consider the case $j = n$.

Theorem 3.8. Let $T \in \operatorname{Sym}^{\tau}(n;\mathcal{O})$ be positive definite, $\varepsilon > 0$ and $k > 2 + 2r(n-1)$, k even for $\mathbb{F} = \mathbb{H}$. Then the POINCARÉ-series $Q^k_{n,n}(Z,T)$ converges absolutely and uniformly in every domain $Z \in \mathcal{V}_{\varepsilon}(n;\mathbb{F})$ and $T \geq \varepsilon I$. $Q^k_{n,n}(\cdot,T)$ represents a cusp form on $H(n;\mathbb{F})$ of weight k and there exists a constant $c = c(k,n,\varepsilon) > 0$ such that

$$\sum_{M : A_{n,n} \backslash \Gamma_n} |e^{2\pi i \tau(T, M\langle Z\rangle)} (\det M\{Z\})^{-k}| \leq c \, (\det Y)^{-\frac{1}{2}k}$$

holds for all $Z = X + iY \in H(n;\mathbb{F})$ and $T \geq \varepsilon I$.

Proof. In view of $T \in \operatorname{Sym}^{\tau}(n;\mathcal{O})$ the definition does not depend on the choice of the representatives. If $T \geq \varepsilon I$, Lemma 3.7 yields

$$\varphi := \sum_{M : A_{n,n} \backslash \Gamma_n} e^{-2\pi\tau(T, \operatorname{Im} M\langle Z\rangle)} \, |\det M\{Z\}|^{-k}$$

$$\le c \sum_{M:A_{n,n}\setminus\Gamma_n} \int_{\mathcal{D}} |\det M\{Z,W\}|^{-k} (\det V)^{\frac{1}{2}k}\, d\nu(W) .$$

Together with M also M^{*-1} runs through a set of representatives of $A_{n,n}\setminus\Gamma_n$. Hence analogous arguments lead to

$$\varphi \le c \sum_{M:A_{n,n}\setminus\Gamma_n} \int_{M^{*-1}<\mathcal{D}>} |\det(Z+W)|^{-k} (\det V)^{\frac{1}{2}k}\, d\nu(W)$$

$$= c\,\gamma \int_{H(n;\mathbb{F})} |\det(Z+W)|^{-k} (\det V)^{\frac{1}{2}k}\, d\nu(W)$$

$$= \alpha\, c\, \gamma\, (\det Y)^{-\frac{1}{2}k} .$$

Proposition 3.1 and Lemma 2.5 yield that φ is uniformly majorized by

$$\sum_{M:A_{n,n}\setminus\Gamma_n} \int_{\mathcal{D}} |\det\{iI,W\}|^{-k} (\det V)^{\frac{1}{2}k}\, d\nu(W) ,$$

whenever $Z \in V_\varepsilon(n;\mathbb{F})$ and $T \ge \varepsilon I$. The same arguments used in the proof of Theorem 3.4 complete the proof.

□

If j is arbitrary, one observes that the map

$$\Gamma_{n,j}\setminus\Gamma_n \times A_{j,j}\setminus\Gamma_j \longrightarrow A_{n,j}\setminus\Gamma_n ,$$

$$(\Gamma_{n,j}M, A_{j,j}K_1) \longmapsto A_{n,j}(K_1 \times I)M ,$$

is bijective. Hence we can proceed as in the proof of Theorem 3.5 in order to derive

<u>Theorem 3.9.</u> <u>Let</u> $1 \le j < n$, $k > 2 + r(n+j-1)$, $k \equiv 0 \bmod s(\mathbb{F})$, $\varepsilon > 0$. <u>Then</u> <u>the</u> POINCARE´-<u>series</u>

$$Q^k_{n,j}(Z,T) = \sum_{M:A_{n,j}\setminus\Gamma_n} e^{2\pi i\tau(T,M<Z>_1)} (\det M\{Z\})^{-k}$$

<u>converges</u> <u>absolutely</u> <u>and</u> <u>uniformly</u> <u>for</u> $Z \in V_\varepsilon(n;\mathbb{F})$ <u>and</u> $T \in \mathrm{Sym}^\tau(j;\mathcal{O})$, $T \ge \varepsilon I$. <u>One</u> <u>has</u>

$$Q^k_{n,j}(\cdot,T) \in [\Gamma(n;\mathbb{F}),k] ,$$

$$E^k_{n,j}(\cdot,Q^k_{j,j}(\cdot,T)) = Q^k_{n,j}(\cdot,T) ,$$

$$Q^k_{n,j}(\cdot,T)\,|\,\phi = Q^k_{n-1,j}(\cdot,T) .$$

On the analogy of Theorem 3.6 we derive

<u>Lemma 3.10.</u> <u>Let</u> $1 \leq j \leq n$, $k > \min\{2 + r(n+j-1), 2 + 2r(n-1)\}$, $k \equiv 0 \bmod s(\mathbb{F})$. <u>Given</u> $T \in \mathrm{Sym}^{\tau}(j;\mathcal{O})$, $T > 0$, $U \in GL(j;\mathcal{O})$ <u>and</u> $Z \in H(n;\mathbb{F})$ <u>one has</u>

$$Q^k_{n,j}(Z,T[U]) = Q^k_{n,j}(Z,T) ,$$

$$Q^k_{n,j}(-\tilde{Z},T) = \overline{Q^k_{n,j}(Z,T)} .$$

<u>The</u> FOURIER-<u>coefficients of</u> $Q^k_{n,j}(\cdot,T)$ <u>turn out to be real numbers</u>.

The connection between the two types of POINCARE´-series is established by

<u>Theorem 3.11.</u> <u>Let</u> $1 \leq j \leq n$, $k > \min\{2 + r(n+j-1), 2 + 2r(n-1)\}$, $k \equiv 0 \bmod s(\mathbb{F})$. <u>Given</u> $Z \in H(n;\mathbb{F})$ <u>and</u> $W \in H(j;\mathbb{F})$ <u>the following</u> FOURIER-<u>expansion</u>

$$\delta(j,k,\mathcal{O})\ P^k_{n,j}(Z,W)$$

$$= \sum_{T \in \mathrm{Sym}^{\tau}(j;\mathcal{O}),\, T>0} (\det T)^{k-1-\frac{1}{2}r(j-1)}\ Q^k_{n,j}(Z,T)\ e^{2\pi i \tau(T,W)}$$

<u>holds</u>, <u>where</u>

$$\delta(j,k,\mathcal{O}) = \mathrm{vol}\ \mathrm{Sym}(j;\mathcal{O})\ (4\pi)^{\frac{1}{4}rj(j-1)}\ (-2\pi i)^{-jk} \prod_{l=0}^{j-1} \Gamma(k-\tfrac{1}{2}rl)$$

<u>Proof.</u> Using Proposition 3.2 one computes that

$$P^k_{n,j}(Z,W) = \sum_{M:A_{n,j}\setminus\Gamma_n} (\det M\{Z\})^{-k} \sum_{S \in \mathrm{Sym}(j;\mathcal{O})} \det(M\langle Z\rangle_1 + W + S)^{-k} .$$

In view of $k > 1 + r(j-1)$ the inner series coincides with

$$\delta(j,k,\mathcal{O})^{-1} \sum_{T \in \mathrm{Sym}^{\tau}(j;\mathcal{O}),\, T>0} (\det T)^{k-1-\frac{1}{2}r(j-1)}\ e^{2\pi i \tau(T, M\langle Z\rangle_1 + W)}$$

according to Lemma 1.7. Interchanging the infinite series completes the proof.

□

§4 Metrization

The generalization of PETERSSON's scalar product to SIEGEL's modular forms was established by MAASS [43],[44]. By means of this metrization principle we can derive various theorems of representation. The assertions with respect to SIEGEL's modular forms are due to MAASS [44] and KLINGEN [32].

We content ourselves with the definition of the scalar product with respect to the full modular group, although we could even consider congruence subgroups according to KOECHER [35].

Let dv denote the symplectic volume element. Given modular forms $f,g \in [\Gamma(n;\mathbb{F}),k]$ we formally define the scalar product of f and g by

$$(f,g) := \int_{\mathcal{F}(n;\mathbb{F})} f(Z)\,\overline{g(Z)}\,(\det Y)^k\,dv(Z)\ .$$

Proposition 4.1. a) The scalar product converges absolutely, whenever f or g is a cusp form.

b) Let $f \in [\Gamma(n;\mathbb{F}),k]_o$ and $g \in [\Gamma(n;\mathbb{F}),k]$ then $(f,g) = \overline{(g,f)}$, $(f,f) \geq 0$ and $(f,f) = 0$ if and only if $f \equiv 0$.

c) $[\Gamma(n;\mathbb{F}),k]_o$ together with the scalar product $(\cdot,\cdot)$ turns out to be a unitary vector space.

Proof. Since $fg \in [\Gamma(n;\mathbb{F}),2k]_o$ the function $|f(Z)|\,|g(Z)|\,(\det Y)^k$ is bounded according to Lemma III.2.4. The symplectic volume of $\mathcal{F}(n;\mathbb{F})$ is finite in view of Proposition II.3.9. Straightforward computations lead to b) and c).

□

Now consider the orthogonal complement of the subspace of cusp forms, where we use the abbreviation $\Gamma_n := \Gamma(n;\mathbb{F})$.

Proposition 4.2. Defining

$$[\Gamma_n,k]_o^{\perp} := \{f \in [\Gamma_n,k]\ ;\ (f,g) = 0 \text{ for all } g \in [\Gamma_n,k]_o\}$$

one has

$$[\Gamma_n,k] = [\Gamma_n,k]_o \oplus [\Gamma_n,k]_o^{\perp}\ .$$

Proof. Obviously $[\Gamma_n,k]_o \cap [\Gamma_n,k]_o^{\perp} = \{0\}$ holds. Given $f \in [\Gamma_n,k]$ then

$$[\Gamma_n,k]_o \longrightarrow \mathbb{C}\ ,\ g \longmapsto (g,f)\ ,$$

turns out to be linear. Since $[\Gamma_n,k]_o$ together with $(\cdot,\cdot)$ is a unitary vector space, there exists an $f_o \in [\Gamma_n,k]_o$ satisfying

$$(g,f) = (g,f_o) \quad \text{for all} \quad g \in [\Gamma_n,k]_o\ .$$

Thus we have $f = f_o + (f-f_o)$, where $f-f_o \in [\Gamma_n,k]_o^{\perp}$.

□

According to MAASS [44],Satz 3 , we introduce subspaces, which are metrically characterized. Put $M_{o,o}^k = [\Gamma(0;\mathbb{F}),k]_o = \mathbb{C}$. Given $n \geq 1$ we define $M_{n,n}^k = [\Gamma(n;\mathbb{F}),k]_o$ and for $0 \leq j < n$

$$M_{n,j}^k := \left\{ f \in [\Gamma(n;\mathbb{F}),k]_o^{\perp}\ ;\ f|\phi \in M_{n-1,j}^k \right\}\ .$$

The surjectivity of SIEGEL's ϕ-operator implies

Theorem 4.3. *Given* $k > 2 + 2r(n-1)$, $k \equiv 0 \bmod s(\mathbb{F})$ *there exists a decomposition*

$$[\Gamma(n;\mathbb{F}),k] = \bigoplus_{j=0}^{n} M_{n,j}^k$$

as direct sums. If $0 \leq j < n$, *the map* $\phi : M_{n,j}^k \longrightarrow M_{n-1,j}^k$ *becomes an isomorphism.*

Proof. We use induction on n , where the case $n = 0$ is trivial and $n = 1$ holds because of Proposition 4.2. In view of Theorem 2.12 the map $\phi : [\Gamma_n,k]_o^{\perp} \longrightarrow [\Gamma_{n-1},k]$ becomes an isomorphism. The induction hypothesis completes the proof.

□

We shall describe the subspaces $M_{n,j}^k$ by means of EISENSTEIN-series. Therefore we need

Lemma 4.4. *Let* $0 \leq j < n$, $k > 2 + r(n+j-1)$, $k \equiv 0 \bmod s(\mathbb{F})$ *and* $f \in [\Gamma(j;\mathbb{F}),k]_o$. *Then*

$$(g, E_{n,j}^k(\cdot,f)) = 0$$

holds for all $g \in [\Gamma(n;\mathbb{F}),k]_o$.

Proof. The definitions yield

$$(g,E^k_{n,j}(\cdot,f))$$

$$= \int_{F_n} \sum_{M:\Gamma_{n,j}\setminus\Gamma_n} g(Z)\ \overline{f(M\langle Z\rangle_1)}\ \overline{(\det M\{Z\})^{-k}}\ (\det Y)^k\ dv(Z) \ ,$$

where $F_n = F(n;\mathbb{F})$. Since g is a cusp form, we obtain that $|g(Z)|(\det Y)^k$ remains bounded in F_n by a slight modification of Lemma III.2.4. As the symplectic volume of F_n is finite and the EISENSTEIN-series converges uniformly in F_n , we may distribute the integral through the infinite series. Thus

$$(g,E^k_{n,j}(\cdot,f))$$

$$= \sum_{M:\Gamma_{n,j}\setminus\Gamma_n} \int_{F_n} \overline{f(M\langle Z\rangle_1)}\ g(Z)\ \overline{(\det M\{Z\})^{-k}}\ (\det Y)^k\ dv(Z)$$

$$= \sum_{M:\Gamma_{n,j}\setminus\Gamma_n} \int_{F_n} \overline{f(M\langle Z\rangle_1)}\ g(M\langle Z\rangle)\ \det(\mathrm{Im}\ M\langle Z\rangle)^k\ dv(Z)$$

$$= \sum_{M:\Gamma_{n,j}\setminus\Gamma_n} \int_{M\langle F_n\rangle} \overline{f(Z_1)}\ g(Z)\ (\det Y)^k\ dv(Z)$$

$$= \int_{F(\Gamma_{n,j})} \overline{f(Z_1)}\ g(Z)\ (\det Y)^k\ dv(Z)$$

where $F(\Gamma_{n,j})$ denotes a fundamental domain of $H(n;\mathbb{F})$ with respect to the action of $\Gamma_{n,j}$ and where the last integral converges absolutely. We obtain from Proposition 2.2 that the integral can be taken over X_4 consisting of the right lower $(n-j)$-rowed block of X , where X_4 runs through $C(n-j;\mathbb{F})$. Since g is a cusp form, the FOURIER-expansion in Lemma III.2.3 yields

$$\int_{C(n-j;\mathbb{F})} g(Z)\ dX_4 = 0 \ .$$

Hence we have

$$(g,E^k_{n,j}(\cdot,f)) = 0 \ . \qquad \square$$

Thus we obtain an assertion on the generators of $[\Gamma(n;\mathbb{F}),k]$ by

<u>Theorem 4.5.</u> <u>Let</u> $0 \leq j \leq n$, $k > 2+2r(n-1)$, $k \equiv 0 \bmod s(\mathbb{F})$. <u>Then one has</u>

$$M^k_{n,j} = \{E^k_{n,j}(\cdot,f)\ ;\ f \in [\Gamma(j;\mathbb{F}),k]_o\}\ ,$$

$$[\Gamma(n;\mathbb{F}),k] = \bigoplus_{j=0}^{n} \{E^k_{n,j}(\cdot,f)\ ;\ f \in [\Gamma(j;\mathbb{F}),k]_o\}\ .$$

<u>Proof.</u> In view of $E^k_{n,n}(\cdot,f) = f$ it suffices to consider $0 \leq j < n$. The preceding lemma yields $E^k_{n,j}(\cdot,f) \in [\Gamma(n;\mathbb{F}),k]^{\perp}_o$. Now the case $n=1$ is trivial and an induction using Theorem 4.3 and Theorem 2.11 completes the proof.

□

If $k > 2+2r(n-1)$ and $k \equiv 0 \bmod s(\mathbb{F})$, especially one has

$$M^k_{n,o} = \mathbb{C}\, E^k_n\ .$$

Thus it suffices to find generators of the subspace of cusp forms. Therefore we consider the POINCARE´-series introduced in §3.

<u>Theorem 4.6.</u> <u>Let</u> $k > 2+2r(n-1)$ <u>and</u> $k \equiv 0 \bmod 2$ <u>for</u> $\mathbb{F} = \mathbb{H}$, $T \in \mathrm{Sym}^{\tau}(n;\mathcal{O})$, $T > 0$. <u>Then every</u> $f \in [\Gamma(n;\mathbb{F}),k]$ <u>fulfills</u>

$$(f, Q^k_{n,n}(\cdot,T)) = \rho\ (\det T)^{-k+h/n}\ \alpha(T)\ ,$$

<u>whenever</u> $\alpha(T)$ <u>denotes the</u> FOURIER-<u>coefficient of</u> T <u>in the</u> FOURIER-<u>expansion of</u> f <u>and</u> $\rho = \gamma\delta$, <u>where</u>

$$\delta = \mathrm{vol}\ \mathrm{Sym}(n;\mathcal{O})\ \pi^{\frac{1}{4}rn(n-1)}\ (4\pi)^{h-kn} \prod_{l=0}^{n-1} \Gamma(k-1-\tfrac{1}{2}r(n+l-1))$$

<u>and</u> $\gamma = \begin{cases} s(\mathbb{F}) & \underline{\text{for}}\ n>1 \\ \mathrm{ord}\ E(\mathbb{F}) & \underline{\text{for}}\ n=1 \end{cases}$.

<u>Proof.</u> In view of the behavior of convergence we may distribute the integral through the infinite series, hence

$$(f, Q^k_{n,n}(\cdot,T))$$

$$= \sum_{M:A_{n,n}\setminus\Gamma_n} \int_{F_n} f(Z)\ e^{-2\pi i\tau(T,\widetilde{M\langle Z\rangle})}\ \overline{(\det M\{Z\})^{-k}}\ (\det Y)^k\, dv(Z)$$

$$= \sum_{M:A_{n,n}\setminus\Gamma_n} \int_{F_n} f(M\langle Z\rangle)\ e^{-2\pi i\tau(T,\widetilde{M\langle Z\rangle})}\ \det(\operatorname{Im} M\langle Z\rangle)^k\ dv(Z)$$

$$= \sum_{M:A_{n,n}\setminus\Gamma_n} \int_{M\langle F_n\rangle} e^{-2\pi i\tau(T,\widetilde{Z})}\ f(Z)\ (\det Y)^k\ dv(Z)\ .$$

By the union $\bigcup_{M:A_{n,n}\setminus\Gamma_n} M\langle F_n\rangle$ a fundamental domain of $H(n;\mathbb{F})$ with respect ot the action of $A_{n,n}$ is covered γ-times. Obviously we can choose $F(A_{n,n}) = C(n;\mathbb{F}) + i\,\operatorname{Pos}(n;\mathbb{F})$ and achieve

$$(f, Q^k_{n,n}(\cdot,T))$$

$$= \gamma \int_{F(A_{n,n})} e^{-2\pi i\tau(T,\widetilde{Z})}\ f(Z)\ (\det Y)^k\ dv(Z)\ .$$

Theorem III.1.2 yields

$$\int_{C(n;\mathbb{F})} f(Z)\ e^{-2\pi i\tau(T,Z)}\ dX = \operatorname{vol}\ \operatorname{Sym}(n;\mathcal{O})\ \alpha(T)\ .$$

In view of Lemma 1.6 the remaining integral

$$\int_{\operatorname{Pos}(n;\mathbb{F})} e^{-4\pi\tau(T,Y)}\ (\det Y)^{k-2h/n}\ dY$$

equals

$$(\det Y)^{-k+h/n}\ \pi^{\frac{1}{4}rn(n-1)}\ (4\pi)^{h-kn} \prod_{l=0}^{n-1} \Gamma(k-1-\tfrac{1}{2}r(n+l-1))\ .$$

□

Let α_T denote the FOURIER-coefficients of $Q^k_{n,n}(\cdot,T)$. Since the α_T 's are real, Proposition 4.1 and the preceding theorem imply

Corollary 4.7. Let $S,T \in \operatorname{Sym}^\tau(n;\mathcal{O})$ be positive definite and $k > 2+2r(n-1)$, $k \equiv 0 \bmod 2$ if $\mathbb{F} = \mathbb{H}$. Then

$$\alpha_S(T)\,(\det T)^{-k+h/n} = \alpha_T(S)\ (\det S)^{-k+h/n}\ .$$

Another consequence using Proposition 4.1 is stated in

Corollary 4.8. Let $k > 2+2r(n-1)$ and $k \equiv 0 \bmod 2$ if $\mathbb{F} = \mathbb{H}$. Then the space of cusp forms $[\Gamma(n;\mathbb{F}),k]_0$ is generated by the POINCARE'-series $Q^k_{n,n}(\cdot,T)$, where $T \in \operatorname{Sym}^\tau(n;\mathcal{O})$ is positive definite.

In view of Theorem III.2.8 it suffices to take those finitely many positive definite $T \in \mathrm{Sym}^{\tau}(n;\mathcal{O})$ satisfying $\mathrm{tr}(T) \le \frac{k}{4\pi}\, s(n;\mathbb{F})$.

Corollary 4.8, Theorem 4.5 and Theorem 3.9 lead to

Corollary 4.9. *Let* $k > 2+2r(n-1)$, $k \equiv 0 \bmod s(\mathbb{F})$. *For* $1 \le j \le n$ *the subspace* $M^k_{n,j}$ *is generated by the* POINCARE´-*series* $Q^k_{n,j}(\cdot,T)$, *where* $T \in \mathrm{Sym}^{\tau}(j;\mathcal{O})$ *is positive definite.*

We are going to derive another theorem of representation. Again let $\gamma = s(\mathbb{F})$, if $n > 1$, and $\gamma = \mathrm{ord}\, E(\mathbb{F})$, if $n = 1$.

Theorem 4.10. *Suppose that* $k > 2+2r(n-1)$, $k \equiv 0 \bmod 2$ *for* $\mathbb{F} = \mathbb{H}$ *and* $W \in H(n;\mathbb{F})$. *Then every cusp form* $f \in [\Gamma(n;\mathbb{F}),k]_0$ *satisfies*

$$(f,P^k_{n,n}(\cdot,W)) = \eta\, f(-\widetilde{W}) ,$$

where

$$\eta = \gamma(2\pi)^h\, 2^n\, (2i)^{-nk} \prod_{l=0}^{n-1} \frac{\Gamma(k-1-\frac{1}{2}r(n+l-1))}{\Gamma(k-\frac{1}{2}rl)} .$$

Proof. Theorem 3.11 yields

$$P^k_{n,n}(Z,W)$$

$$= \delta^{-1} \sum_{T\in\mathrm{Sym}^{\tau}(n;\mathcal{O}),T>0} (\det T)^{k-h/n}\, Q^k_{n,n}(Z,T)\, e^{2\pi i\tau(T,W)} ,$$

where $\delta = \mathrm{vol}\,\mathrm{Sym}(n;\mathcal{O})\, (4\pi)^{\frac{1}{4}rn(n-1)}\, (-2\pi i)^{-nk} \prod_{l=0}^{n-1} \Gamma(k-\frac{1}{2}rl)$. In view of the uniform convergence in vertical strips we may distribute the integral through the infinite series. Since f is a cusp form, f possesses a FOURIER-expansion of the form

$$f(Z) = \sum_{T\in\mathrm{Sym}^{\tau}(n;\mathcal{O}),T>0} \alpha(T)\, e^{2\pi i\tau(T,Z)} .$$

We obtain from Theorem 4.6 that

$$(f,P^k_{n,n}(\cdot,W))$$

$$= \delta^{-1} \sum_{T\in\mathrm{Sym}^{\tau}(n;\mathcal{O}),T>0} (\det T)^{k-h/n}\, (f,Q^k_{n,n}(\cdot,T))\, e^{-2\pi i\tau(T,\widetilde{W})}$$

$$= \delta^{-1} \rho \sum_{T \in \mathrm{Sym}^{\tau}(n;\mathcal{O}),\, T>0} \alpha(T)\, e^{-2\pi i \tau(T,\widetilde{W})}$$

$$= \delta^{-1} \rho\, f(-\widetilde{W}) \; .$$

□

Thus we achieve a representation by an integral

Corollary 4.11. *Let* $k > 2 + 2r(n-1)$ *and* $k \equiv 0 \bmod 2$ *for* $\mathbb{F} = \mathbb{H}$. *Then every cusp form* $f \in [\Gamma(n;\mathbb{F}),k]_0$ *satisfies*

$$f(W) = \xi \int_{H(n;\mathbb{F})} f(Z)\, \det(W-\widetilde{Z})^{-k}\, (\det Y)^k\, dv(Z) \; , \quad W \in H(n;\mathbb{F}) \; ,$$

where

$$\xi = (2\pi)^{-h}\, 2^{-n}\, (2i)^{nk} \prod_{l=0}^{n-1} \frac{\Gamma(k-\frac{1}{2}rl)}{\Gamma(k-1-\frac{1}{2}r(n+l-1))} \; .$$

Proof. Because of the behavior of convergence we may distribute the integral through the infinite series and Theorem 4.10 yields

$$f(W) = \eta^{-1} \int_{F_n} f(Z)\, \overline{P^k_{n,n}(Z,-\widetilde{W})}\, (\det Y)^k\, dv(Z)$$

$$= \eta^{-1} \int_{F_n} \sum_{M \in \Gamma_n} f(M\langle Z\rangle)\, \det(\widetilde{M\langle Z\rangle}-W)^{-k} \det(\operatorname{Im} M\langle Z\rangle)^k\, dv(Z)$$

$$= \eta^{-1} \sum_{M \in \Gamma_n} \int_{M\langle F_n\rangle} f(Z)\, \det(\widetilde{Z}-W)^{-k}\, (\det Y)^k\, dv(Z)$$

$$= \xi \int_{H(n;\mathbb{F})} f(Z)\, \det(W-\widetilde{Z})^{-k}\, (\det Y)^k\, dv(Z) \; ,$$

where $\xi = \gamma\eta^{-1}$.

□

On the analogy of Corollary 4.8 and Corollary 4.9 one can prove

Corollary 4.12. *Let* $k > 2 + 2r(n-1)$ *and* $k \equiv 0 \bmod 2$ *for* $\mathbb{F} = \mathbb{H}$. *Then the space of cusp forms* $[\Gamma(n;\mathbb{F}),k]_0$ *is generated by the* POINCARÉ-*series* $P^k_{n,n}(\cdot,W)$, *where* $W \in H(n;\mathbb{F})$.

<u>Corollary 4.13.</u> <u>Let</u> $k > 2 + 2r(n-1)$ <u>and</u> $k \equiv 0 \bmod s(\mathbb{F})$. <u>Given</u> $1 \leq j \leq n$ <u>the subspace</u> $M^k_{n,j}$ <u>is generated by the</u> POINCARE´-<u>series</u> $P^k_{n,j}(\cdot,W)$, <u>where</u> $W \in H(j;\mathbb{F})$.

We add a final theorem of representation.

<u>Lemma 4.14.</u> <u>Let</u> $1 \leq j \leq n$, $k > 2 + r(n+j-1)$ <u>and</u> $k \equiv 0 \bmod s(\mathbb{F})$, $Z \in H(n;\mathbb{F})$. <u>Then all cusp forms</u> $f \in [\Gamma(j;\mathbb{F}),k]_0$ <u>satisfy</u>

$$(f,P^k_{n,j}(Z,\cdot)) = \eta\, E^k_{n,j}(-\widetilde{Z},f) .$$

<u>The space of cusp forms</u> $[\Gamma(j;\mathbb{F}),k]_0$ <u>is generated by the</u> POINCARE´-<u>series</u> $P^k_{n,j}(Z,\cdot)$, <u>where</u> $Z \in H(n;\mathbb{F})$.

<u>Proof.</u> Given $Z \in H(n;\mathbb{F})$ and $W \in H(j;\mathbb{F})$ Theorem 3.5 and Theorem 3.6 yield

$$P^k_{n,j}(Z,W) = E^k_{n,j}(Z,P^k_{j,j}(W,\cdot)) .$$

The same arguments as used above lead to

$$\begin{aligned}(f,P^k_{n,j}(Z,\cdot)) &= \sum_{M:\Gamma_{n,j}\backslash\Gamma_n} (f,P^k_{j,j}(\cdot,M\langle Z\rangle_1))\,\overline{(\det M\{Z\})^{-k}} \\ &= \eta \sum_{M:\Gamma_{n,j}\backslash\Gamma_n} f(-\widetilde{M\langle Z\rangle_1})\,\overline{(\det M\{Z\})^{-k}} \\ &= \eta \sum_{M:\Gamma_{n,j}\backslash\Gamma_n} f(M^*\langle-\widetilde{Z}\rangle_1)\,(\det M^*\{-\widetilde{Z}\})^{-k} \\ &= \eta\, E^k_{n,j}(-\widetilde{Z},f) ,\end{aligned}$$

if we regard Proposition 3.1, since even M^* runs through a set of representatives of $\Gamma_{n,j}\backslash\Gamma_n$. Given a cusp form $f \in [\Gamma(j;\mathbb{F}),k]_0$ satisfying $(f,P^k_{n,j}(Z,\cdot)) = 0$ for all $Z \in H(n;\mathbb{F})$ one has $E^k_{n,j}(\cdot,f) \equiv 0$. We obtain $f \equiv 0$ from Theorem 2.11.

□

§5 Algebraical independence

Theorem III.2.12 says that every $h+2$ modular forms on $H(n;\mathbb{F})$ are algebraically dependent. By means of POINCARÉ-series we shall construct $h+1$ algebraically independent cusp forms of a common weight. According to the classic procedure by CARTAN [11],[12] and PYATECKIǏ-ŠAPIRO [46] we show that inequivalent points of the half-space can be separated by cusp forms. The representation leans on [21],I,§4, and [13],Kap.V .

Two points $Z,W \in H(n;\mathbb{F})$ are called <u>equivalent modulo</u> $\Gamma(n;\mathbb{F})$ if there exists $M \in \Gamma(n;\mathbb{F})$ satisfying $M\langle Z\rangle = W$, otherwise they are said to be <u>inequivalent modulo</u> $\Gamma(n;\mathbb{F})$.

A point $Z \in H(n;\mathbb{F})$ is called a <u>fixed point of the modular group</u> $\Gamma(n;\mathbb{F})$ if there exists an $M \in \Gamma(n;\mathbb{F})$ such that the corresponding modular transformation is different from the identity and satisfies $M\langle Z\rangle = Z$.

Given $Z \in H(n;\mathbb{F})$ by the stabilizer of Z we mean

$$\mathrm{Stab}(Z) := \mathrm{Stab}(Z;\mathbb{F}) := \{M \in \Gamma(n;\mathbb{F}) \; ; \; M\langle Z\rangle = Z\} .$$

The subgroup $\mathrm{Stab}(Z)$ of $\Gamma(n;\mathbb{F})$ turns out to be finite in view of Corollary II.3.8. According to Corollary I.5.7 there exists $t = t(n;\mathbb{F})$ satisfying

(1) $$\mathrm{ord}\,\mathrm{Stab}(Z) \,|\, t$$

for all $Z \in H(n;\mathbb{F})$. Let this t be fixed. Clearly

$$t \equiv 0 \bmod s(\mathbb{F}) .$$

We obtain

(2) $$(M\{Z\})^k = M^k\{Z\} \; , \; (M\{Z\})^t = I$$

from Theorem II.1.7 and (1), whenever $M \in \mathrm{Stab}(Z)$ and $k \in \mathbb{Z}$.

Throughout this paragraph let $z_1,\ldots,z_h$ denote the independent complex coordinates of a matrix $Z \in H(n;\mathbb{F})$ in an arbitrary, but fixed order.

<u>Proposition 5.1.</u> <u>Let</u> $Z_1,\ldots,Z_m$ <u>be mutually distinct points in</u> $H(n;\mathbb{F})$ <u>and</u> $\alpha_j,\beta_{jk} \in \mathbb{C}$ <u>for</u> $1 \le j \le m$, $1 \le k \le h$. <u>Then there exists a bounded holomorphic function</u> $\varphi : H(n;\mathbb{F}) \longrightarrow \mathbb{C}$ <u>satisfying</u>

$$\varphi(Z_j) = \alpha_j \quad \text{and} \quad \left(\frac{\partial\varphi}{\partial z_k}\right)_{Z=Z_j} = \beta_{jk}$$

<u>for</u> $1 \leq j \leq m$, $1 \leq k \leq h$.

<u>Proof.</u> Consider $\hat{Z} \in H(rn;\mathbb{R})$ and transform SIEGEL's half-space onto the generalized unit circle (cf. SIEGEL [55],§4). Thus it suffices to find a polynomial in h variables, which besides its partial derivatives takes given values in m mutually distinct points. The solution of this problem is well-known.

□

Given $\varphi : H(n;\mathbb{F}) \longrightarrow \mathbb{C}$ and $Z \in H(n;\mathbb{F})$ we use the abbreviation

$$P_n^k(Z,\varphi) := P_{n,n}^k(Z,iI,\varphi) .$$

$P_n^k(\cdot,\varphi)$ turns out to be a cusp form of weight k, whenever φ is holomorphic and bounded and $k > 2 + 2r(n-1)$.

Following CHRISTIAN [13],Satz 5.66, we prove

<u>Theorem 5.2.</u> <u>Let</u> $Z_1,\ldots,Z_m \in H(n;\mathbb{F})$. <u>Then there exists</u> $g_0 \in \mathbb{N}$ <u>such that for all</u> $g \in \mathbb{N}$, $g \geq g_0$ <u>there is a cusp form</u> $f \in [\Gamma(n;\mathbb{F}),tg]_0$ <u>satisfying</u>

$$f(Z_j) \neq 0 , \quad 1 \leq j \leq m .$$

<u>Proof.</u> Without restriction we may suppose that the orbits of $Z_1,\ldots,Z_m$ are mutually disjoint, since $f \in [\Gamma_n,k]$, $f(Z) \neq 0$ and $M \in \Gamma_n$ imply

$$f(M\langle Z\rangle) = (\det M\{Z\})^k f(Z) \neq 0 .$$

One has $t > 2 + 2r(n-1)$. According to Lemma 3.3 the series

$$\sum_{M\in\Gamma_n} |\det M\{Z,iI\}|^{-t} , \quad Z \in H(n;\mathbb{F}) ,$$

converges. Hence there are only finitely many $M_1,\ldots,M_l \in \Gamma_n$, where $M_1 = I$, such that

$$\sum{}' \, |\det M\{Z_j,iI\}|^{-t} \leq 2^{-t} \, |\det(Z_j+iI)|^{-t} \quad \text{for } 1 \leq j \leq m ,$$

where the dash may indicate that the sum is taken over $M \in \Gamma_n$, $M \neq M_1,\ldots,M_l$. Thus we have

$$\sum{}' \, |\det M\{Z_j,iI\}|^{-tg} \leq 2^{-tg} \, |\det(Z_j+iI)|^{-tg} \tag{3}$$

for $1 \leq j \leq m$ and all $g \geq 1$. According to Proposition 5.1 there exists a holomorphic function $\varphi : H(n;\mathbb{F}) \longrightarrow \mathbb{C}$ satisfying $|\varphi(Z)| \leq c$, $\varphi(Z_j) = 1$ and $\varphi(M_k<Z_j>) = 0$ for $1 \leq k \leq l$, $1 \leq j \leq m$, whenever $M_k \notin \mathrm{Stab}(Z_j)$. Now (2) yields

$$\left|\sum_{k=1}^{l} \varphi(M_k<Z_j>) \ (\det M_k\{Z_j,iI\})^{-tg}\right| \geq |\det(Z_j+iI)|^{-tg} .$$

Choosing g_o such that $2^{-tg_o}c < 1$ we calculate from (3) for $g \geq g_o$ and $1 \leq j \leq m$ that

$$\begin{aligned} \left|P_n^{tg}(Z_j,\varphi)\right| &\geq \left|\sum_{k=1}^{l} \varphi(M_k<Z_j>) \ (\det M_k\{Z_j,iI\})^{-tg}\right| \\ &\quad - c \sum{}' |\det M\{Z_j,iI\}|^{-tg} \\ &\geq (1 - c2^{-tg})|\det(Z_j+iI)|^{-tg} > 0 . \end{aligned}$$

□

In view of Theorem 4.10 an analogous argument leads to

<u>Corollary 5.3.</u> <u>Let</u> $1 \leq j \leq n$, $W \in H(j;\mathbb{F})$. <u>Then there exists a</u> $k_o = k_o(W)$ <u>such that the</u> POINCARE´-<u>series</u> $P_{n,j}^{k}(\cdot,W)$ <u>does not vanish identically, whenever</u> $k \geq k_o(W)$ <u>and</u> $k \equiv 0 \bmod \mathrm{ord}\,\mathrm{Stab}(W)$.

On the analogy of [21],I.4.4 , we derive

<u>Lemma 5.4.</u> <u>Let</u> $Z_1,Z_2 \in H(n;\mathbb{F})$ <u>be inequivalent modulo</u> $\Gamma(n;\mathbb{F})$. <u>Then there exists</u> $g_o \in \mathbb{N}$ <u>such that for all</u> $g \in \mathbb{N}$, $g \geq g_o$, <u>there is a cusp form</u> $f \in [\Gamma(n;\mathbb{F}),tg]_o$ <u>satisfying</u>

$$|f(Z_1)| > |\det(Z_1+iI)|^{-tg} \quad \underline{\text{and}} \quad |f(Z_2)| < |\det(Z_2+iI)|^{-tg} .$$

<u>Proof.</u> Choose the same notations as in the proof of Theorem 5.2 and let (3) hold. According to Proposition 5.1 we determine a holomorphic function $\varphi : H(n;\mathbb{F}) \longrightarrow \mathbb{C}$ satisfying $|\varphi(Z)| \leq c$, $\varphi(Z_1) = 2$, $\varphi(M_k<Z_1>) = 0$, whenever $1 \leq k \leq l$ and $M_k \notin \mathrm{Stab}(Z_1)$, as well as $\varphi(M_k<Z_2>) = 0$ for $1 \leq k \leq l$. Since Z_1 and Z_2 are inequivalent, these condition can be fulfilled. Now choose g_o satisfying $2^{-tg_o}c < 1$. Analogous arguments lead to

$$|P_n^{tg}(Z_1,\varphi)| \geq (2 - c2^{-tg}) \ |\det(Z_1+iI)|^{-tg} > |\det(Z_1+iI)|^{-tg} ,$$

$$|P_n^{tg}(Z_2,\varphi)| = \left| \sum{}' \varphi(M\langle Z_2\rangle)(\det M\{Z_2,iI\})^{-tg} \right|$$

$$\leq c\, 2^{-tg}\, |\det(Z_2+iI)|^{-tg}$$

$$< |\det(Z_2+iI)|^{-tg} .$$

□

Thus we can separate inequivalent points by cusp forms.

<u>Theorem 5.5.</u> <u>Let</u> $Z_1, Z_2 \in H(n;\mathbb{F})$ <u>be</u> <u>inequivalent</u> <u>modulo</u> $\Gamma(n;\mathbb{F})$. <u>There</u> <u>exists</u> $k_o > 0$ <u>such</u> <u>that</u> <u>for</u> <u>all</u> $k \geq k_o$, $k \equiv 0 \bmod t$, <u>there</u> <u>is</u> <u>a</u> <u>cusp</u> <u>form</u> $f \in [\Gamma(n;\mathbb{F}),k]_o$ <u>satisfying</u>

$$f(Z_1) \neq 0 \quad \underline{\text{and}} \quad f(Z_2) = 0 .$$

<u>Proof.</u> According to the preceding lemma there exists g_o such that for all $g \geq g_o$ one can find cusp forms $f_1, f_2 \in [\Gamma(n;\mathbb{F}),tg]_o$ satisfying

$$|f_1(Z_1)| > |(\det(Z_1+iI)|^{-tg} , \quad |f_1(Z_2)| < |\det(Z_2+iI)|^{-tg} ,$$

$$|f_2(Z_1)| < |(\det(Z_1+iI)|^{-tg} , \quad |f_2(Z_2)| > |\det(Z_2+iI)|^{-tg} .$$

Put $f(Z) := f_1(Z)\, f_2(Z_2) - f_2(Z)\, f_1(Z_2)$, then $f(Z_1) \neq 0$ and $f(Z_2) = 0$.

□

Next we want to construct $h+1$ algebraically independent modular forms. Therefore we use the idea of [13], Satz 6.5, in order to prove

<u>Lemma 5.6.</u> <u>Given</u> $Z_o \in H(n;\mathbb{F})$ <u>there</u> <u>exists</u> $c = c(Z_o) > 0$ <u>such</u> <u>that</u>

$$\left| \frac{\partial(\det M\{Z,iI\})^{-2}}{\partial z_k} \right|_{Z=Z_o} \leq c\, |\det M\{Z_o,iI\}|^{-2}$$

<u>holds</u> <u>for</u> $1 \leq k \leq h$ <u>and</u> <u>all</u> $M \in Sp(n;\mathbb{F})$.

<u>Proof.</u> First choose a compact neighborhood $\bar{U}$ of Z_o in $H(n;\mathbb{F})$. Let the point Z differ von Z_o only in the k-th coordinate. Putting $w := z_k - (Z_o)_k$ then Z belongs to U whenever $|w| < 2\varepsilon$ for some $\varepsilon > 0$. Then $\psi(w) := (\det M\{Z,iI\})^{-2}$ turns out to be a holomorphic function in the disk $\{w \in \mathbb{C} ;\ |w| < 2\varepsilon\}$. CAUCHY's integral formula

yields

$$\psi'(0) = \frac{1}{2\pi i} \int_{|w|=\varepsilon} \psi(w)\, w^{-2}\, dw ,$$

$$|\psi'(0)| \le \varepsilon^{-1} \max_{|w|=\varepsilon} |\psi(w)| ,$$

$$\left| \frac{\partial (\det M\{Z,iI\})^{-2}}{\partial z_k} \right| \le \varepsilon^{-1} \max_{W\in\bar{U}} |\det M\{W,iI\}|^{-2} .$$

In view of Proposition 3.1 and Lemma 2.5 there exists $c = c(\bar{U})$ such that

$$|\det M\{W,iI\}|^{-2} \le c\, |\det M\{Z_0,iI\}|^{-2}$$

for all $W \in \bar{U}$ and $M \in Sp(n;\mathbb{F})$.

□

The knowledge of the growth of the partial derivatives is used to prove

<u>Theorem 5.7.</u> <u>Given a point</u> $Z_0 \in H(n;\mathbb{F})$, <u>which is not a fixed point of the modular group, there exist cusp forms</u> $f_0,\ldots,f_h$ <u>of a common weight such that the</u> WRONSKI-<u>determinant in</u> Z_0

$$W(f_0,\ldots,f_h)_{Z=Z_0} = \det \begin{pmatrix} f_0 & \cdots\cdots & f_h \\ \frac{\partial f_0}{\partial z_1} & & \frac{\partial f_h}{\partial z_1} \\ \vdots & & \vdots \\ \frac{\partial f_0}{\partial z_h} & \cdots\cdots & \frac{\partial f_h}{\partial z_h} \end{pmatrix}_{Z = Z_0}$$

<u>does not vanish</u>.

<u>Proof.</u> In case $n = 1$ it suffices to consider $\mathbb{F} = \mathbb{R}$. Hence one has $\mathrm{Stab}(Z_0) = \{\varepsilon I \,;\, \varepsilon \in S(\mathbb{F})\}$, because Z_0 is not a fixed point of $\Gamma(n;\mathbb{F})$. On the analogy of the proof of Theorem 5.2 there exist matrices $M_1 = I$, $M_2,\ldots,M_l$ such that

$$(4) \qquad \sum_{M\in\mathcal{M}} |\det M\{Z_0,iI\}|^{-t} \le (2c)^{-t}\, |\det(Z_0+iI)|^{-t} ,$$

where $c = c(Z_o) > 1$ is determined according to Lemma 5.6 and $\mathcal{M} = \Gamma_n - \{\varepsilon M_j \; ; \; \varepsilon \in S(\mathbb{F}) \; , \; 1 \leq j \leq l\}$. Next Lemma 5.6 yields

$$\sum_{M\in\mathcal{M}} \left|\frac{\partial(\det M\{Z,iI\})^{-tg}}{\partial z_k}\right|_{Z=Z_o} \leq \tfrac{1}{2}\, c t g\, (2c)^{-tg}\, |\det(Z_o+iI)|^{-tg}$$

for $g \geq 1$ and $1 \leq k \leq h$. Hence we obtain

$$(5) \qquad \sum_{M\in\mathcal{M}} \left|\frac{\partial(\det M\{Z,iI\})^{-tg}}{\partial z_k}\right|_{Z=Z_o} \leq 2^{-tg}\, |\det(Z_o+iI)|^{-tg}$$

for all $g \geq g_o$. Since Z_o is not a fixed point, the points $M_j\langle Z_o\rangle$, $1 \leq j \leq l$, are mutually distinct. In view of Proposition 5.1 there exist holomorphic functions $\varphi_m : H(n;\mathbb{F}) \longrightarrow \mathbb{C}$, $0 \leq m \leq h$, such that

$$(6) \qquad \begin{cases} |\varphi_m(Z)| \leq C \,, & 0 \leq m \leq h \,, \\ \varphi_o(Z_o) = 1 \,,\; \varphi_o(M_j\langle Z_o\rangle) = 0 \,, & 2 \leq j \leq l \,, \\ \varphi_m(M_j\langle Z_o\rangle) = 0 \,, & 1 \leq m \leq h \,,\; 1 \leq j \leq l \,, \\ \left(\dfrac{\partial\varphi_m(Z)}{\partial z_k}\right)_{Z=M_j\langle Z_o\rangle} = \delta_{j1}\delta_{km} \,, & 1 \leq j \leq l \,,\; 0 \leq m \leq h \,,\; 1 \leq k \leq h \,. \end{cases}$$

Thus we have

$$\left(\frac{\partial\varphi_m(M_j\langle Z\rangle)}{\partial z_k}\right)_{Z=Z_o} = 0 \,,\; 2 \leq j \leq l \,.$$

In view of CAUCHY's integral formula there exists $\varepsilon = \varepsilon(Z_o) > 0$ such that

$$(7) \qquad \left|\frac{\partial\varphi_m(M\langle Z\rangle)}{\partial z_k}\right|_{Z=Z_o} \leq \varepsilon^{-1}\, C$$

holds for $M \in Sp(n;\mathbb{F})$, $0 \leq m \leq h$, $1 \leq k \leq h$. One has

$$f_m^g(Z) := P_n^{tg}(Z,\varphi_m) \in [\Gamma(n;\mathbb{F}),tg]_o \,.$$

(4) and (6) yield

$$\lim_{g\to\infty} f_m^g(Z_o) = \delta_{om}\, s \,,\quad 0 \leq m \leq h \,.$$

Moreover, (5),(6) and (7) imply

$$\lim_{g\to\infty} \left(\frac{\partial f_m^g(Z)}{\partial z_k}\right)_{Z=Z_o} = \delta_{km}\, s \,,\; 0 \leq m \leq h \,,\; 1 \leq k \leq h \,,$$

hence

$$\lim_{g\to\infty} W(f_o^g,\ldots,f_n^g)_{Z=Z_o} = s^{h+1} \neq 0 \,.$$

□

According to [21],I.4.7, we derive

Theorem 5.8. There exist $h+1$ algebraically independent cusp forms of a suitable common weight.

Proof. Choose a point Z_0, which is not a fixed point of $\Gamma(n;\mathbb{F})$. In virtue of Theorem 5.7 there exist cusp forms $f_0,\dots,f_h$ of a common weight satisfying $W(f_0,\dots,f_h)_{Z=Z_0} \neq 0$. Since $f_j(Z_0) = 0$ for $0 \leq j \leq h$ contradicts $W(f_0,\dots,f_h)_{Z=Z_0} \neq 0$, we may suppose without restriction that $f_0(Z_0) \neq 0$. Hence there exists an open neighborhood $\mathcal{U}$ of Z_0 in $H(n;\mathbb{F})$ such that $f_0(Z) \neq 0$ for all $Z \in \mathcal{U}$. Thus the functions

$$g_j : \mathcal{U} \longrightarrow \mathbb{C}\ ,\ g_j(Z) := f_0(Z)^{-1} f_j(Z) \quad , 1 \leq j \leq h \ ,$$

are holomorphic. The JACOBIan determinant

$$\det\left(\frac{\partial g_j(Z)}{\partial z_k}\right)_{Z=Z_0} = f_0(Z_0)^{-h-1}\, W(f_0,\dots,f_h)_{Z=Z_0}$$

does not equal 0. Hence $(g_1,\dots,g_h)$ map an open neighborhood of Z_0 biholomorphically onto an open subset V in $\mathbb{C}^h$.

Suppose that $f_0,\dots,f_h$ are algebraically dependent. Then there is $m \in \mathbb{N}$ such that the monomials $f_0^{m_0}\cdot\ldots\cdot f_h^{m_h}$, $m_0 + \dots + m_h = m$, and hence even $g_1^{m_1}\cdot\ldots\cdot g_h^{m_h}$, $m_1 + \dots + m_h \leq m$, are linearly dependent. Given a polynomial $p \in \mathbb{C}[X_1,\dots,X_h]$ satisfying $p(g_1,\dots,g_h) \equiv 0$ then p vanishes identically on V, thus $p \equiv 0$ and we have a contradiction.

□

Analogous arguments lead to

Theorem 5.9. a) There exist n^2+1 cusp form $f_0,\dots,f_{n^2}$ in $[\Gamma(n;\mathbb{H}),k]_0$ of a common weight $k \equiv 0 \bmod 4$ such that $f_0\big|_{H(n;\mathbb{C})},\dots,f_{n^2}\big|_{H(n;\mathbb{C})}$ are algebraically independent symmetric Hermitian modular forms.

b) There exist $\frac{1}{2}n(n+1)+1$ cusp forms $f_0,\dots,f_m$, $m = \frac{1}{2}n(n+1)$, in $[\Gamma(n;\mathbb{F}),k]_0$ of a common weight such that $f_0\big|_{H(n;\mathbb{R})},\dots,f_m\big|_{H(n;\mathbb{R})}$ become algebraically independent SIEGEL's modular forms.

<u>Proof.</u> We may suppose $n > 1$.

a) Determine $Z_o \in H(n;\mathbb{C}) \subset H(n;\mathbb{H})$ such that $\mathrm{Stab}(Z_o;\mathbb{H})$ equals εI , $\varepsilon = \pm e_1, \pm e_2$. Let $z_1,\dots,z_m$, $m = n^2$, denote the independent variables of Z in $H(n;\mathbb{C}) \subset H(n;\mathbb{H})$. Just in the same way, as we did in the proof of Theorem 5.7, we can show that there exist cusp forms $f_o,\dots,f_m$ in $[\Gamma(n;\mathbb{H}),k]_o$ of a suitable common weight $k \equiv 0 \bmod 4$ such that

$$\det \begin{pmatrix} f_o & \cdots\cdots & f_m \\ \frac{\partial f_o}{\partial z_1} & & \frac{\partial f_m}{\partial z_1} \\ \cdot & & \cdot \\ \cdot & & \cdot \\ \cdot & & \cdot \\ \frac{\partial f_o}{\partial z_m} & \cdots\cdots & \frac{\partial f_m}{\partial z_m} \end{pmatrix}_{Z=Z_o} \neq 0 .$$

Theorem III.4.1 yields that $f_o|H(n;\mathbb{C}),\dots,f_m|H(n;\mathbb{C})$ are symmetric Hermitian cusp forms of weight k . On the analogy of Theorem 5.8 we conclude that they are algebraically independent.

b) Proceed in the same way, where one has to choose $Z_o \in H(n;\mathbb{R}) \subset H(n;\mathbb{F})$ satisfying $\mathrm{Stab}(Z_o;\mathbb{F}) = \{\varepsilon I \ ; \ \varepsilon \in E(\mathbb{F})\}$.

□

The assertions above will be applied to modular functions in the next chapter in order to determine the transcendence degree of the field of modular functions.

Chapter VI Modular functions

Modular functions are defined to be meromorphic functions on the half-space, which remain invariant under modular transformations and in addition turn out to be meromorphic at ∞ , whenever $n = 1$.

First SATAKE [51] and BAILY [3] used the compactification of SIEGEL's half-space and deep results of the general complex analysis in order to show that SIEGEL's modular functions can be represented as quotients of SIEGEL's modular forms of the same weight. Later on SIEGEL [57] gave a classic proof of this fact. In the sequel we use the paper of ANDREOTTI and GRAUERT [1]. Thus the field of modular functions on the half-space $H(n;\mathbb{F})$ turns out to be an algebraic function field of transcendence degree h .

We apply these results to modular functions with respect to congruence subgroups. Restricting the domain of definition modular functions of quaternions and Hermitian modular functions turn into SIEGEL's modular functions, whenever they are not singular on the whole submanifold. By means of a result of the theory of compactification of SIEGEL's half-space and by the virtue of the property of separating points one can show that each SIEGEL's modular function becomes the restriction of a Hermitian modular function resp. of a modular function of quaternions. I.e. we can imbed the field of SIEGEL's modular functions into the field of Hermitian modular functions as well as into the field of modular functions of quaternions.

§1 The field of modular functions

In this paragraph we examine the field of modular functions, determine its degree of transcendency and show that each modular function can be represented as a quotient of two modular forms of the same weight.

Definition. A function f is called meromorphic on $H(n;\mathbb{F})$ if f is defined on an open dense subset $\mathcal{U}$ of $H(n;\mathbb{F})$ and if for every $Z_0 \in H(n;\mathbb{F})$ there exist an open neighborhood $\mathcal{U}(Z_0) \subset H(n;\mathbb{F})$ as well as holomorphic functions φ,ψ on $\mathcal{U}(Z_0)$ such that $\psi(Z) \neq 0$ and $f(Z) = \frac{\varphi(Z)}{\psi(Z)}$ for all $Z \in \mathcal{U}(Z_0) \cap \mathcal{U}$.

Definition. A function f is said to be a modular function on $H(n;\mathbb{F})$ if the following conditions are satisfied:

(MF.1) f is meromorphic on $H(n;\mathbb{F})$.

(MF.2) $f(M\langle Z\rangle) = f(Z)$ for all $M \in \Gamma(n;\mathbb{F})$ and $Z \in H(n;\mathbb{F})$.

(MF.3) f does not possess an essential singularity at ∞, whenever $n = 1$.

The set of all modular functions on $H(n;\mathbb{F})$ is denoted by

$$M_n := M(\Gamma(n;\mathbb{F})) .$$

Modular functions on $H(n;\mathbb{R})$ resp. $H(n;\mathbb{C})$ resp. $H(n;\mathbb{H})$ are called SIEGEL's modular functions resp. Hermitian modular functions resp. modular functions of quaternions.

Clearly, M_n contains the constant maps and therefore forms an extension field of $\mathbb{C}$.

$\Gamma(1;\mathbb{F})$ induces the same transformations on the upper half in $\mathbb{C}$, whenever $\mathbb{F} = \mathbb{R},\mathbb{C},\mathbb{H}$. Hence M_1 coincides with the field of elliptic modular functions.

Theorem 1.1. a) Let $f,g \in [\Gamma(n;\mathbb{F}),k]$, where $g \neq 0$. Then $\frac{f}{g}$ is a modular function on $H(n;\mathbb{F})$.

b) There are h (over $\mathbb{C}$) algebraically independent modular functions on $H(n;\mathbb{F})$, which can be represented as quotients of modular forms of the same weight.

c) Given $Z,Z_0 \in H(n;\mathbb{F})$, which are inequivalent modulo $\Gamma(n;\mathbb{F})$,

<u>there exists a modular function</u> f <u>on</u> $H(n;\mathbb{F})$ <u>satisfying</u>

$$f(Z) \neq f(Z_o) .$$

<u>Proof.</u> a) The assertion is clear, since (MF.3) follows from (M.3).

b) According to Theorem V.5.8 there exist $h+1$ algebraically independent cusp forms $f_o,\dots,f_h$ of a common weight. Hence $f_o \not\equiv 0$ and $\frac{f_1}{f_o},\dots,\frac{f_h}{f_o}$ form algebraically independent modular functions.

c) Use Theorem V.5.5 and Theorem V.5.2.

□

We need another result on the behavior of modular transformations. Let $\Sigma_{n,j}$ resp. $\Gamma_{n,j}$ again denote the subgroups of Σ_n examined in Proposition V.2.1. Remember the definition of SIEGEL's elementary set $S(n;\mathbb{F})[\alpha]$, $\alpha > 0$, in II, §3. Now Satz 5.24 in [13] is formulated by

<u>Lemma 1.2.</u> <u>Given</u> $\alpha > 0$ <u>there exists</u> $c(\alpha) > 0$ <u>such that one has for all</u> $M \in \Gamma(n;\mathbb{F})$, $M \notin \bigcup_{j=0}^{n-1} \Gamma_{n,j}$:

$$S(n;\mathbb{F})[\alpha] \cap M\langle\{Z \in S(n;\mathbb{F})[\alpha] \; ; \; \det Y \geq c(\alpha)\}\rangle = \emptyset .$$

<u>Proof.</u> Without restriction assume $\alpha > 1$, hence $\hat{S}(n;\mathbb{F})[\alpha] \subset S(rn;\mathbb{R})[\alpha]$. By the virtue of Corollary II.3.8 there exist only finitely many $M \in \Gamma(n;\mathbb{F})$ satisfying $M\langle S(n;\mathbb{F})[\alpha]\rangle \cap S(n;\mathbb{F})[\alpha] \neq \emptyset$. We obtain $\hat{M} \notin \bigcup_{j=0}^{rn-1} \Sigma_{rn,j}$ from $M \notin \bigcup_{j=0}^{n-1} \Gamma_{n,j}$. In view of [13], Satz 4.110, the diagonal elements and hence the determinant of V remain bounded, whenever

$$U+iV \in \hat{M}\langle S(rn;\mathbb{R})[\alpha]\rangle \cap S(rn;\mathbb{R})[\alpha] .$$

But $\det \hat{Y} = (\det Y)^r$ for $Y \in Pos(n;\mathbb{F})$ completes the proof.

□

Next we state some definitions from [1].

Definition. Let $\mathcal{U} \subset \mathbb{C}^h$ be a region and G a subgroup of Bih $\mathcal{U}$.

a) If $V \subset \mathcal{U}$ is open then

$$V^{\mathcal{U}} := \left\{ z \in \mathcal{U} \;;\; \begin{array}{l} |f(z)| \leq \sup |f(V)| \quad \text{for all holomorphic} \\ \text{maps} \quad f : \mathcal{U} \longrightarrow \mathbb{C} \end{array} \right\}$$

is called the holomorphic convex hull of V with respect to $\mathcal{U}$.

b) Let $V \subset \mathcal{U}$ be open and $z \in \partial V$ be a boundary point. z is said to be a pseudoconcave boundary point of V if there exist arbitrarily small neighborhoods $\mathcal{W} = \mathcal{W}(z) \subset \mathbb{C}^h$ such that z becomes an interior point of $(\mathcal{W} \cap V)^{\mathcal{W}}$.

c) G is called pseudoconcave if there exists an open relatively compact subset $V \subset \mathcal{U}$ such that the topological closure $\overline{V}$ is contained in $\mathcal{U}$ and every boundary point $z \in \partial V$ can be mapped onto an interior point or onto a pseudoconcave boundary point of V by an element of G.

Given $n > 1$ the following matrix will be used in the sequel

$$(1) \qquad W = \begin{pmatrix} 0 & \cdots & 0 & w \\ \vdots & & & 0 \\ \vdots & & & \vdots \\ 0 & & 0 & 0 \\ w & 0 \cdots & 0 & w_n \end{pmatrix} \in \mathrm{Sym}(n;\mathbb{R}) \otimes_{\mathbb{R}} \mathbb{C} \subset \mathrm{Sym}(n;\mathbb{F}) \otimes_{\mathbb{R}} \mathbb{C} ,$$

where $W = U+iV$ and $w = u+iv$ as well as $w_n = u_n + iv_n$ are complex variables. Next we state a suitable version of a criterion for pseudoconcave boundary points (cf. [1], Satz 1).

Lemma 1.3. Let $n > 1$, $V \subset \mathrm{Sym}(n;\mathbb{F}) \otimes_{\mathbb{R}} \mathbb{C}$ be a region and Z a boundary point of V. Z is a pseudoconcave boundary point of V if there exist $\varepsilon > 0$, $\mathcal{U}_\varepsilon := \{Z + W \;;\; W$ of the form (1), $|w| + |w_n| < \varepsilon\}$ and a real-valued twice continuously differentiable function $\varphi = \varphi(w,w_n,\tilde{w},\tilde{w}_n)$ on $\mathcal{U}_\varepsilon$ such that the following holds:

a) the Hermitian form

$$H(\varphi) = \frac{\partial^2\varphi}{\partial w \partial \tilde{w}} dw d\tilde{w} + \frac{\partial^2\varphi}{\partial w \partial \tilde{w}_n} dw d\tilde{w}_n + \frac{\partial^2\varphi}{\partial w_n \partial \tilde{w}} dw_n d\tilde{w} + \frac{\partial^2\varphi}{\partial w_n \partial \tilde{w}_n} dw_n d\tilde{w}_n$$

is positive definite in $\mathcal{U}_\varepsilon$,

b)
$$\mathcal{U}_\varepsilon \cap V = \{Z + W \;;\; \varphi(w,w_n,\tilde{w},\tilde{w}_n) > 0\} ,$$
$$\mathcal{U}_\varepsilon \cap \partial V = \{Z + W \;;\; \varphi(w,w_n,\tilde{w},\tilde{w}_n) = 0\} .$$

Consider the function

$$\psi : H(n;\mathbb{F}) \longrightarrow \mathbb{R} , \ \psi(Z) := -\log(\det Y) .$$

In view of the criterion above we first derive

<u>Proposition 1.4.</u> <u>Let</u> $n > 1$, $Z \in H(n;\mathbb{F})$ <u>and</u> $\varepsilon > 0$ <u>such that</u> $\mathcal{U}_\varepsilon \subset H(n;\mathbb{F})$. <u>Then the Hermitian form</u> $H(\psi)$ <u>is positive definite in</u> $\mathcal{U}_\varepsilon$.

<u>Proof.</u> One easily calculates that

$$H(\psi) = \frac{\partial^2\psi}{\partial v^2}(dv^2 + du^2) + \frac{\partial^2\psi}{\partial v_n^2}(dv_n^2 + du_n^2) + 2\frac{\partial^2\psi}{\partial v\partial v_n}(dvdv_n + dudu_n).$$

Let $Y + V = \begin{pmatrix} Y_1 & \mathfrak{y} \\ \bar{\mathfrak{y}}' & y_n+v_n \end{pmatrix}$, where $Y_1 \in \mathrm{Pos}(n-1;\mathbb{F})$, $\mathfrak{y} \in \mathbb{F}^{n-1}$. Now $Y+V > 0$ leads to $\Delta = y_n + v_n - Y_1^{-1}[\mathfrak{y}] > 0$ and $\det Y = \Delta \det Y_1$ in view of Lemma I.3.2. Putting $Y_1^{-1} = (y^*_{kl})$ one easily checks that

$$\frac{\partial^2\psi}{\partial v_n^2} = \Delta^{-2} ,$$

$$\frac{\partial^2\psi}{\partial v\partial v_n} = \Delta^{-2} \frac{\partial\Delta}{\partial v} ,$$

$$\frac{\partial^2\psi}{\partial v^2} = \Delta^{-2} (2y_1^*\Delta + (\frac{\partial\Delta}{\partial v})^2) .$$

By the virtue of Theorem I.3.11 $H(\psi)$ is positive definite because of $y_1^* > 0$.

□

Thus we have all the tools in order to derive

<u>Lemma 1.5.</u> <u>Given</u> $n > 1$ <u>the group of modular transformations</u>

$$\{Z \longmapsto M\langle Z\rangle \ ; \ M \in \Gamma(n;\mathbb{F})\}$$

<u>is pseudoconcave</u>.

<u>Proof.</u> Let $M \in Sp(n;\mathbb{F})$, then one has $\psi(M\langle Z\rangle) = \psi(Z) + \log|\det M\{Z\}|^2$ in view of Theorem II.1.7. We obtain from the definition of $\mathcal{F}(n;\mathbb{F})$ that ψ takes its minimum on each equivalence class in a point be-

longing to $F(n;\mathbb{F})$. Hence

$$\varphi : H(n;\mathbb{F}) \longrightarrow \mathbb{R} \ , \ \varphi(Z) := \min \{\psi(M\langle Z\rangle) \ ; \ M \in \Gamma(n;\mathbb{F})\} \ ,$$

is well-defined. According to Proposition II.3.6 there exists $\alpha = \alpha(n;\mathbb{F}) > 0$ satisfying $F(n;\mathbb{F}) \subset S(n;\mathbb{F})[\alpha]$. Hence φ is continuous in $S(n;\mathbb{F})[\alpha]$. Applying Lemma 1.2 we determine $c = c(2\alpha) > 0$. By means of Corollary II.3.8 there exist only finitely many $M \in \Gamma(n;\mathbb{F})$ such that $M\langle S(n;\mathbb{F})[2\alpha]\rangle \cap S(n;\mathbb{F})[2\alpha] \neq \emptyset$, among these let $M_1,\ldots,M_t$ belong to $\bigcup_{j=0}^{n-1} \Gamma_{n,j}$. Defining

$$T := \{Z \in S(n;\mathbb{F})[\alpha] \ ; \ \det Y > c\}$$

Lemma 1.2 yields

$$\varphi(Z) = \min \{\psi(M_k\langle Z\rangle) \ ; \ 1 \leq k \leq t\} \quad \text{for} \quad Z \in T \ .$$

Given $M \in \bigcup_{j=0}^{n-1} \Gamma_{n,j}$ then $|\det M\{Z\}|^2$ does neither depend on the last row nor on the last column of Z . Given $Z \in T$ and $\varepsilon > 0$ such that $U_\varepsilon \subset T$ then $H(\psi)$ is positive definite in U_ε according to Proposition 1.4.

Now choose $V := \{Z \in S(n;\mathbb{F})[\alpha] \ ; \det Y < c+1\}$. Clearly V is an open relatively compact subset of $H(n;\mathbb{F})$ such that its topological closure also is contained in $H(n;\mathbb{F})$. Every $Z \in \partial V$ satisfying $\det Y = c+1$ is a pseudoconcave boundary point in view of Lemma 1.3, where we have to replace the function φ defined above by $\varphi + \log(c+1)$. The other boundary points of V are equivalent modulo $\Gamma(n;\mathbb{F})$ to interior points of V .

Now we can apply the results of ANDREOTTI and GRAUERT [1].

<u>Theorem 1.6.</u> <u>The field</u> $M(\Gamma(n;\mathbb{F}))$ <u>has the transcendence degree</u> $h = n + \frac{1}{2}r\,n(n-1)$ <u>over</u> $\mathbb{C}$.

<u>There are modular functions</u> $f_o,\ldots,f_h \in M(\Gamma(n;\mathbb{F}))$ <u>such that</u>

$$M(\Gamma(n;\mathbb{F})) = \mathbb{C}(f_o,\ldots,f_h) \ .$$

<u>If</u> $f_1,\ldots,f_h \in M(\Gamma(n;\mathbb{F}))$ <u>are algebraically independent the field extension</u> $M(\Gamma(n;\mathbb{F})) \supset \mathbb{C}(f_1,\ldots,f_h)$ <u>is finite and there exists</u> $f \in M(\Gamma(n;\mathbb{F}))$ <u>satisfying</u>

$$M(\Gamma(n;\mathbb{F})) = \mathbb{C}(f,f_1,\ldots,f_h) .$$

Every modular function is a quotient of two modular forms of the same weight.

Proof. In the case $n = 1$ confer [52], p.89 . In the case $n > 1$ there exist h algebraically independent modular functions according to Theorem 1.1. Thus Satz 3 and Satz 4 in [1] complete the proof.

□

In the well-known case $n = 1$ of elliptic modular functions there exists a more detailed description. Remembering the definition of the EISENSTEIN-series E_n^k in V, §2, we put

$$j := \frac{(E_1^4)^3}{(E_1^4)^3 - (E_1^6)^2} .$$

Then one has

$$M_1 = \mathbb{C}(j)$$

(cf. [52], p.89).

§2 Modular functions with respect to congruence subgroups and symmetric modular functions

On the analogy of modular forms we can extend the definition of modular functions to congruence subgroups.

Definition. Let C be a congruence subgroup of $\Gamma(n;\mathbb{F})$. A function f is called a modular function on $H(n;\mathbb{F})$ with respect to the congruence subgroup C if the following properties are satisfied:

(MC.1) f is meromorphic on $H(n;\mathbb{F})$.

(MC.2) $f(M\langle Z\rangle) = f(Z)$ for all $Z \in H(n;\mathbb{F})$ and $M \in C$.

(MC.3) If $n = 1$ then $f \underset{0}{|} M$ does not possess an essential singularity at ∞, whenever $M \in SL(2;\mathbb{Q})$.

The set of modular functions with respect to C is denoted by $M(C)$.

Obviously $M(C)$ is an extension field of $M(\Gamma(n;\mathbb{F}))$, since in the case $C = \Gamma(n;\mathbb{F})$ both definitions coincide. On the analogy of [35], Satz 6, we derive

Theorem 2.1. Let C_1 and C_2 be two congruence subgroups of $\Gamma(n;\mathbb{F})$ satisfying $C_1 \subset C_2$. Then $M(C_1) \supset M(C_2)$ is a field extension, where

$$[M(C_1):M(C_2)] \leq [C_2:C_1] .$$

Proof. Since C_1 and C_2 are congruence subgroups, the index $k := [C_2:C_1]$ is finite. Let $M_1,\ldots,M_k$ denote a set of representatives of the right cosets, i.e. $C_2 = \bigcup_{j=1}^{k} C_1 M_j$. Given $f \in M(C_1)$ define the polynomial

$$\varphi(X;Z) := \prod_{j=1}^{k} (X - f(M_j\langle Z\rangle)) , \quad Z \in H(n;\mathbb{F}) .$$

By construction one has

$$\varphi(X;M\langle Z\rangle) = \varphi(X;Z)$$

for all $M \in C_2$ and $Z \in H(n;\mathbb{F})$. Therefore we obtain

$$\varphi(X;\cdot) = X^k + g_{k-1}X^{k-1} + \ldots + g_0 ,$$

where $g_j \in M(C_2)$. Because of $\varphi(f(Z);Z) = 0$ for all $Z \in H(n;\mathbb{F})$ the degree of each $f \in M(C_1)$ over $M(C_2)$ does not exceed k . Choosing $f \in M(C_1)$ possessing maximal degree over $M(C_2)$ leads to $M(C_1) = M(C_2)[f]$ and hence to

$$[M(C_1):M(C_2)] \leq k .$$

□

<u>Corollary 2.2.</u> <u>Let</u> $C \subset \Gamma(n;\mathbb{F})$ <u>be a congruence subgroup. Then the transcendence degree of</u> $M(C)$ <u>over</u> $\mathbb{C}$ <u>equals</u> h .
<u>There exist</u> $f_0,\ldots,f_h \in M(C)$ <u>satisfying</u>

$$M(C) = \mathbb{C}(f_0,\ldots,f_h) .$$

<u>Every modular function with respect to</u> C <u>can be represented as the quotient of two modular forms with respect to</u> C <u>and of the same weight.</u>

<u>Proof.</u> Use Theorem 2.1 and Theorem 1.6. □

On the analogy of modular forms we can define <u>symmetric modular functions</u> by the property $f(Z) = f(Z')$ for all $Z \in H(n;\mathbb{F})$, whenever $\mathbb{F} = \mathbb{C}$, $n \geq 2$, or $\mathbb{F} = \mathbb{H}$, $n = 2$. The field of symmetric modular functions is denoted by $M^s(\Gamma(n;\mathbb{F}))$.

<u>Proposition 2.3.</u> <u>Let</u> $\mathbb{F} = \mathbb{C}$, $n \geq 2$, <u>or</u> $\mathbb{F} = \mathbb{H}$, $n = 2$. <u>Given</u> $f \in M(\Gamma(n;\mathbb{F}))$ <u>then</u> $f_0(Z) = f(Z)\, f(Z')$ <u>and</u> $f_1(Z) := f(Z) + f(Z')$ <u>are symmetric modular functions.</u>

<u>Proof.</u> It suffices to show that f_0 and f_1 are invariant under modular transformations. Given $M \in \Gamma(n;\mathbb{F})$ we determine $M^* \in \Gamma(n;\mathbb{F})$ such that $(M\langle Z\rangle)' = M^*\langle Z'\rangle$ for all $Z \in H(n;\mathbb{F})$. In the case $\mathbb{F} = \mathbb{C}$ choose $M^* := \bar{M} \in \Gamma(n;\mathbb{C})$.

Therefore let $\mathbb{F} = \mathbb{H}$, $n = 2$. It suffices to consider all the matrices M belonging to the set of generators of $\Gamma(2;\mathbb{H})$ quoted in Theorem II.2.3. Given $M = \begin{pmatrix} I & S \\ O & I \end{pmatrix}$, $S \in Sym(2;\mathcal{O})$, choose $M^* = \begin{pmatrix} I & S' \\ O & I \end{pmatrix}$.

One easily checks $J^* = J$. Considering $M = \begin{pmatrix} \bar{U}' & 0 \\ 0 & U^{-1} \end{pmatrix}$, $U \in GL(2;\mathcal{O})$, we choose $M^* = \begin{pmatrix} \bar{U}^{*\prime} & 0 \\ 0 & U^{*-1} \end{pmatrix}$, $U^* \in GL(2;\mathcal{O})$. It suffices to consider those U belonging to the set of generators of $GL(2;\mathcal{O})$ quoted in Theorem I.2.3. If $U = \begin{pmatrix} 0 & 1 \\ 1 & 0 \end{pmatrix}$ or $U = \begin{pmatrix} 1 & 1 \\ 0 & 1 \end{pmatrix}$ choose $U^* = U$. Let $U = [\varepsilon, 1]$, $\varepsilon \in E$, then put $U^* = [1,\varepsilon]$.

□

Thus we can prove

Theorem 2.4. Let $\mathbb{F} = \mathbb{C}$, $n \geq 2$, or $\mathbb{F} = \mathbb{H}$, $n = 2$, then

$$[M(\Gamma(n;\mathbb{F})) : M^s(\Gamma(n;\mathbb{F}))] = 2 .$$

Proof. Given $f \in M(\Gamma(n;\mathbb{F}))$ then f is a zero of the polynomial $X^2 - f_1 X + f_0$ in $M^s(\Gamma(n;\mathbb{F}))[X]$ according to the preceding proposition. Hence one has $[M_n : M_n^s] \leq 2$ and it suffices to demonstrate the existence of a modular function being non-symmetric.

In view of Theorem 1.1 it suffices to show that there exists $Z \in H(n;\mathbb{F})$ such that Z and Z' are not equivalent modulo $\Gamma(n;\mathbb{F})$. In the case $\mathbb{F} = \mathbb{H}$ consider iY , where $Y = \begin{pmatrix} 15 & y \\ \bar{y} & 16 \end{pmatrix}$, $y = 7e_1 + 3e_2 + 2e_3 + e_4$. If $\mathbb{F} = \mathbb{C}$ we can restrict ourselves to $n = 2$. In this case consider iY , where $Y = \begin{pmatrix} 5 & y \\ \bar{y} & 6 \end{pmatrix}$, $y = 2e_1 + e_2$. Elementary computations show that iY and iY' are not equivalent modulo $\Gamma(2;\mathbb{F})$.

□

Symmetric Hermitian modular functions arise if we restrict modular functions of quaternions to the Hermitian half-space.

§3 Relations between SIEGEL modular functions, Hermitian modular functions and modular functions of quaternions

We consider SIEGEL's and Hermitian half-space as analytic submanifolds of the half-space of quaternions. Restricting the domain of definition modular functions, which are not singular on the whole submanifold, turn into modular functions again. Therefore let $M(\Gamma(n;\mathbb{F}))_{\mathbb{R}}$, $\mathbb{F} = \mathbb{C},\mathbb{H}$, resp. $M(\Gamma(n;\mathbb{H}))_{\mathbb{C}}$ denote the subring of those modular functions on $H(n;\mathbb{F})$ resp. $H(n;\mathbb{H})$, which are not singular in all the points of $H(n;\mathbb{R})$ resp. $H(n;\mathbb{C})$. Thus the maps

$$\sigma : M(\Gamma(n;\mathbb{F}))_{\mathbb{R}} \longrightarrow M(\Gamma(n;\mathbb{R})), \quad f \longmapsto f|_{H(n;\mathbb{R})},$$

$$\sigma : M(\Gamma(n;\mathbb{H}))_{\mathbb{C}} \longrightarrow M(\Gamma(n;\mathbb{C})), \quad f \longmapsto f|_{H(n;\mathbb{C})},$$

are homomorphisms of the rings and $\sigma(M(\Gamma(n;\mathbb{F}))_{\mathbb{R}})$ resp. $\sigma(M(\Gamma(n;\mathbb{H}))_{\mathbb{C}})$ become subfields of the corresponding fields of SIEGEL's resp. Hermitian modular functions.

In the case $n = 1$ SIEGEL's, Hermitian and modular functions of quaternions coincide.

__Lemma 3.1.__ __Let__ $n \geq 2$, __then the field extensions__
$M^s(\Gamma(n;\mathbb{C})) \supset \sigma(M(\Gamma(n;\mathbb{H}))_{\mathbb{C}})$ __and__ $M(\Gamma(n;\mathbb{R})) \supset \sigma(M(\Gamma(n;\mathbb{F}))_{\mathbb{R}})$ __are finite__.

__Proof.__ Put $M = e_3 I \in \Gamma(n;\mathbb{H})$. Given $Z \in H(n;\mathbb{C})$ one has $M\langle Z\rangle = Z'$. Hence each element of $\sigma(M(\Gamma(n;\mathbb{H}))_{\mathbb{C}})$ turns out to be a symmetric Hermitian modular function. By analogy with the proof of Theorem 1.1 we obtain from Theorem V.5.9 that there exist at least n^2 algebraically independent modular functions in $\sigma(M(\Gamma(n;\mathbb{H}))_{\mathbb{C}})$ and at least $\frac{1}{2}n(n+1)$ in $\sigma(M(\Gamma(n;\mathbb{F}))_{\mathbb{R}})$. According to Theorem 1.6 the field extensions under consideration are finite.

□

A more detailed description seems not to be possible if we use the available means.

Using the compactification of SIEGEL's half-space KLINGEN [30] showed that each SIEGEL's modular function can be obtained by the restriction of a Hermitian modular function with respect to an arbitrary imaginary-quadratic number field. FREITAG [18] generalized this result. We use the same method to embed the field of SIEGEL's modular functions into the field of modular functions of quaternions.

Definition. Let T be a locally compact topological space having a denumerable base. A closed subset $A \subset T$ is called thin if the interior of A is empty. A subset $B \subset T$ is said to be thick if the complement of B is contained in a denumberable union of thin subsets.

Especially, the intersection of denumberably many thick subsets becomes thick again. According to BAIRE's principle of categories every thick subset turns out to be dense.

From FREITAG [21],A.3.4, we adopt

Lemma 3.2. Let $F \subset M(\Gamma(n;\mathbb{R}))$ be a field, which contains $\frac{1}{2}n(n+1)$ algebraically independent modular functions. Let a thick subset $U \subset H(n;\mathbb{R})$ exist such that given modulo $\Gamma(n;\mathbb{R})$ inequivalent points $Z, Z_0 \in U$ there is an $f \in F$, which is holomorphic in a neighborhood of U and satisfies $f(Z) \neq f(Z_0)$. Then one has

$$F = M(\Gamma(n;\mathbb{R})) .$$

Considering symmetric Hermitian modular functions an analogous result ought to hold. But in order to prove the corresponding lemma we would have to deduce some results from the theory of compactification or we would have to extract these results from the general theory by BAILY and BOREL [4]. This would digress from the subject of this volume too much.

Next we construct a suitable subset of $H(n;\mathbb{R})$ having the property of separating points. Therefore we have to study those biholomorphic maps from $H(n;\mathbb{F})$ onto itself, which map SIEGEL's half-space onto itself.

Proposition 3.3. Given $n \in \mathbb{N}$ the set

$$\Omega(n;\mathbb{F}) := \{M \in Sp(n;\mathbb{F}) \ ; \ M\langle H(n;\mathbb{R})\rangle = H(n;\mathbb{R})\}$$

consists of all matrices εM, where $M \in Sp(n;\mathbb{R})$ and $\varepsilon \in \mathbb{F}$, $|\varepsilon| = 1$.

Proof. According to Theorem II.1.8 the map

$$\Omega(n;\mathbb{F}) \longrightarrow \mathrm{Bih}\, H(n;\mathbb{R}) \ , \ M \longmapsto (Z \longmapsto M\langle Z\rangle) \ ,$$

turns out to be a surjective homomorphism of the groups. Now one easily verifies that the kernel of this map consists of the matrices εI, where $\varepsilon \in \mathbb{F}$, $|\varepsilon| = 1$.

□

$\Omega(n;\mathbb{F})$ induces the same transformations on $H(n;\mathbb{R})$ as $Sp(n;\mathbb{R})$ does. Thus we achieve

Lemma 3.4. Given $M \in Sp(n;\mathbb{F})$, which does not belong to $\Omega(n;\mathbb{F})$, then the set $M\langle H(n;\mathbb{R})\rangle \cap H(n;\mathbb{R})$ is thin in $H(n;\mathbb{R})$.

Proof. Let $M \in Sp(n;\mathbb{F})$ and $\mathcal{U}$ be an open subset of $H(n;\mathbb{R})$ satisfying $M\langle \mathcal{U}\rangle \subset H(n;\mathbb{R})$. In order to show that $M \in \Omega(n;\mathbb{F})$ holds we may multiply M by elements of $\Omega(n;\mathbb{F})$ from the right and from the left. Hence we may suppose that $\mathcal{U}$ is a neighborhood of iI. Let $iS \in \mathcal{U}$, where $S = [s_1,\ldots,s_n]$ and $1 < s_1 < \ldots < s_n$. According to SIEGEL [55], Lemma 2, we can choose M such that $M\langle iI\rangle = iI$ and $M\langle iS\rangle = iT$, where $T = [t_1,\ldots,t_n]$ and $1 \leq t_1 \leq \ldots \leq t_n$.

We obtain $M = \begin{pmatrix} A & B \\ -B & A \end{pmatrix} \in Sp(n;\mathbb{F})$ from $M\langle iI\rangle = iI$. Now $S > I$, $T \geq I$ and $M\langle iS\rangle = iT$ yield $B = 0$, hence $S[\bar{A}'] = T$, where $A\bar{A}' = \bar{A}'A = I$. From $1 < s_1 < \ldots < s_n$ and $1 \leq t_1 \leq \ldots \leq t_n$ we conclude that A is diagonal. Thus we have $A = [\varepsilon_1,\ldots,\varepsilon_n]$, where $|\varepsilon_j| = 1$, $1 \leq j \leq n$. If $n > 1$ it follows that $\varepsilon_j x \bar{\varepsilon}_{j+1}$ becomes real again, whenever x is real and sufficiently small. Thus we have $M = \varepsilon I$, $\varepsilon \in \mathbb{F}$, $|\varepsilon| = 1$.

□

Let $\Lambda(n;\mathbb{F}) := \Gamma(n;\mathbb{F}) \cap \Omega(n;\mathbb{F})$, hence

$$\Lambda(n;\mathbb{F}) = \{\varepsilon M \ ; \ M \in \Gamma(n;\mathbb{R}) \ ; \ \varepsilon \in E(\mathbb{F})\} \ .$$

Thus $\Lambda(n;\mathbb{F})$ induces the same transformations on $H(n;\mathbb{R})$ as SIEGEL's modular group $\Gamma(n;\mathbb{R})$ does. Hence we can prove the final

<u>Theorem 3.5.</u> <u>Given</u> SIEGEL'<u>s</u> <u>modular function</u> $f \in \mathcal{M}(\Gamma(n;\mathbb{R}))$ <u>there exists a modular function</u> $g \in \mathcal{M}(\Gamma(n;\mathbb{F}))$, $\mathbb{F} = \mathbb{C},\mathbb{H}$, <u>such that</u>

$$g|_{H(n;\mathbb{R})} = f ,$$

<u>i.e.</u> $\sigma(\mathcal{M}(\Gamma(n;\mathbb{F}))_{\mathbb{R}}) = \mathcal{M}(\Gamma(n;\mathbb{R}))$.

<u>Proof.</u> In the case $n = 1$ the assertion is trivial. Let $n > 1$ and $F := \sigma(\mathcal{M}(\Gamma(n;\mathbb{F}))_{\mathbb{R}})$. Lemma 3.1 yields that the transcendence degree of F equals $\frac{1}{2}n(n+1)$. According to Lemma 3.2 it suffices to find a thick subset $\mathcal{U}$ of $H(n;\mathbb{R})$ having the property of separating points. According to Lemma 3.4 the subset

$$\mathcal{U}_o := \bigcap_{\substack{M \in \Gamma(n;\mathbb{F}) \\ M \notin \Lambda(n;\mathbb{F})}} M\langle H(n;\mathbb{R})\rangle^c \cap H(n;\mathbb{R})$$

becomes thick in $H(n;\mathbb{R})$. $\mathcal{U}$ arises from $\mathcal{U}_o$ by removing the zeros of the modular form $\Theta(\cdot,S_{\mathbb{R}};\mathbb{R}) \in [\Gamma(n;\mathbb{R}),4]$ (cf. Corollary IV.2.7). Hence $\mathcal{U}$ forms a thick subset of $H(n;\mathbb{R})$.

Let $Z,Z_o \in \mathcal{U}$ be inequivalent modulo $\Gamma(n;\mathbb{R})$. Hence Z,Z_o are inequivalent modulo $\Gamma(n;\mathbb{F})$. According to Theorem V.5.5 there exist $k > 0$ and $f \in [\Gamma(n;\mathbb{F}),4k]_o$ satisfying $f(Z_o) = 0$ and $f(Z) \neq 0$. Now $g := f \cdot (\Theta(\cdot,S_{\mathbb{F}};\mathbb{F}))^{-k}$ belongs to $\mathcal{M}(\Gamma(n;\mathbb{F}))$. According to Corollary IV.2.7 the restricted function $g|_{H(n;\mathbb{R})}$ turns out to be holomorphic in a neighborhood of $\mathcal{U}$ and satisfies $g(Z_o) = 0$ and $g(Z) \neq 0$. Finally Lemma 3.2 yields

$$F = \mathcal{M}(\Gamma(n;\mathbb{R})) .$$

□

Bibliography

[1] ANDREOTTI, A., and GRAUERT, H.: Algebraische Körper von automorphen Funktionen. Nachr. Akad. Wiss. Göttingen 1961, 39-48.

[2] ANDRIANOV, A.N., and MALOLETKIN, G.N.: Behaviour of theta-series of degree N under modular substitutions. Math. USSR Izvestija 9, 227-241 (1975).

[3] BAILY, W.L.: Satake's compactification of V_n. Amer. J. Math. 80, 348-364 (1958).

[4] BAILY, W.L., and BOREL, A.: Compactification of arithmetic quotients of bounded symmetric domains. Ann. of Math., II.Ser., 84, 442-528 (1966).

[5] BÖCHERER, S.: Über die Fourier-Jacobi-Entwicklung Siegelscher Eisensteinreihen. Math. Z. 183, 21-46 (1983).

[6] BRAUN, H.: Konvergenz verallgemeinerter Eisensteinscher Reihen. Math. Z. 44, 387-397 (1939).

[7] BRAUN, H.: Hermitian Modular Functions I. Ann. of Math., II.Ser., 50, 827-855 (1949). II. Ann. of Math., II.Ser., 51, 92-104 (1950). III. Ann. of Math., II.Ser., 53, 143-160 (1951).

[8] BRAUN, H.: Der Basissatz für hermitische Modulformen. Abh. Math. Sem. Univ. Hamburg 19, 134-148 (1955).

[9] BRAUN, H., and KOECHER, M.: Jordan-Algebren. Grundlehren der mathematischen Wissenschaften 128. Berlin-Heidelberg-New York: Springer (1966).

[10] CARLSSON, R., and JOHANSSEN, W.: Der Multiplikator von Thetareihen höheren Grades zu quadratischen Formen ungerader Ordnung. Math. Z. 177, 439-447 (1981).

[11] CARTAN, H.: Fonctions automorphes et espaces analytique. Séminaire No.6, E.N.S. 1953/54.

[12] CARTAN, H.: Fonctions automorphes et séries de Poincaré. J. d' Analyse Math. 6, 169-175 (1958); Coll. Works II, 731-737.

[13] CHRISTIAN, U.: Siegelsche Modulfunktionen. 2. Auflage. Vorlesungsausarbeitung, Göttingen (1981).

[14] COHN, P.M.: Free rings and their relations. London-New York: Academic Press (1971).

[15] EICHLER, M.: Einführung in die Theorie der algebraischen Zahlen und Funktionen. Basel-Stuttgart: Birkhäuser (1963).

[16] EICHLER, M.: Über die Anzahl der linear unabhängigen Siegelschen Modulformen von gegebenem Gewicht. Math. Ann. 213, 281-291 (1975). Erratum. Math. Ann. 215, 195 (1975).

[17] FREITAG, E.: Zur Theorie der Modulformen zweiten Grades. Nachr. Akad. Wiss. Göttingen 1965, 151-157.

[18] FREITAG, E.: Fortsetzung von automorphen Funktionen. Math. Ann. 177, 95-100 (1968).

[19] FREITAG, E.: Holomorphe Differentialformen zu Kongruenzgruppen der Siegelschen Modulgruppe. Inventiones Math. 30, 181-196 (1975).

[20] FREITAG. E.: Stabile Modulformen. Math. Ann. 230, 197-211 (1977).

[21] FREITAG, E.: Siegelsche Modulfunktionen. Grundlehren der mathematischen Wissenschaften 254. Berlin-Heidelberg-New York: Springer (1983).

[22] HELGASON, S.: Differential Geometry and Symmetric Spaces. New York-San Francisco-London: Academic Press (1962).

[23] HERTNECK, C.: Positivitätsbereiche und Jordan-Strukturen. Dissertation, Münster (1959).

[24] HIRZEBRUCH, U.: Halbräume und ihre holomorphen Automorphismen. Math. Ann. 153, 395-417 (1964).

[25] HURWITZ, A.: Vorlesungen über die Zahlentheorie der Quaternionen. Berlin: Springer (1919).

[26] IGUSA, J.: On Siegel modular forms of genus two I. Amer. J. Math. 84, 175-200 (1962). II. Amer. J. Math. 86, 392-412 (1964).

[27] JACOBSON, N.: Basic Algebra I. San Francisco: Freeman (1974).

[28] KLINGEN, H.: Zur Theorie der hermitischen Modulfunktionen. Math. Ann. 134, 355-384 (1958).

[29] KLINGEN, H.: Quotientendarstellung Hermitescher Modulfunktionen durch Modulformen. Math. Ann. 143, 1-18 (1961).

[30] KLINGEN, H.: Über einen Zusammenhang zwischen Siegelschen und Hermiteschen Modulfunktionen. Abh. Math. Sem. Univ. Hamburg 27, 1-12 (1964).

[31] KLINGEN, H.: Über Poincarésche Reihen zur Siegelschen Modulgruppe. Math. Ann. 168, 157-170 (1967).

[32] KLINGEN, H.: Zum Darstellungssatz für Siegelsche Modulformen. Math. Z. 102, 30-43 (1967). Berichtigung. Math. Z. 105, 399-400 (1968).

[33] KLINGEN, H.: Über Poincarésche Reihen vom Exponentialtyp. Math. Ann. 234, 145-157 (1978).

[34] KOECHER, M.: Über Dirichlet-Reihen mit Funktionalgleichung. J. Reine Angew. Math. 192, 1-23 (1953).

[35] KOECHER, M.: Zur Theorie der Modulformen n-ten Grades I. Math. Z. 59, 399-416 (1954). II. Math. Z. 61, 455-466 (1955).

[36] KOECHER, M.: Beiträge zu einer Reduktionstheorie in Positivitätsbereichen I. Math. Ann. 141, 384-432 (1960).

[37] KOECHER, M.: Matrizen-Rechnung und Modulformen. Vorlesungsausarbeitung, Münster (1980).

[38] KOECHER, M.: Lineare Algebra und analytische Geometrie. 2. Auflage. Grundwissen Mathematik 2. Berlin-Heidelberg-New York-Tokyo: Springer (1985).

[39] KRAZER, A.: Lehrbuch der Thetafunktionen. Leipzig: Teubner (1903). Reprint. New York: Chelsea (1970).

[40] KRIEG, A.: Modulfunktionen auf dem Quaternionen-Halbraum. Dissertation, Münster (1983).

[41] KRIEG, A.: Hecke-operators with respect to the modular group of quaternions. In preparation.

[42] KRONECKER, L.: Über bilineare Formen. J. Reine Angew. Math. 68, 273-285 (1868); Werke I, 143-162.

[43] MAASS, H.: Über die Darstellung der Modulformen n-ten Grades durch Poincarésche Reihen. Math. Ann. 123, 125-151 (1951).

[44] MAASS, H.: Die Primzahlen in der Theorie der Siegelschen Modulfunktionen. Math. Ann. 124, 87-122 (1951).

[45] MAASS, H.: Siegel's Modular Forms and Dirichlet Series. Lecture Notes in Mathematics 216. Berlin-Heidelberg-New York: Springer (1971).

[46] PYATECKIǏ-ŠAPIRO, I.I.: Singular modular functions. Amer. Math. Soc. Transl., II.Ser., 10, 13-58 (1958).

[47] QUEBBEMANN, H.-G.: An application of Siegel's formula over quaternion orders to the existence of extremal lattices. Preprint.

[48] RESNIKOFF, H.L.: On a class of linear differential equations for automorphic forms in several complex variables. Amer. J. Math. 95, 321-332 (1973).

[49] RESNIKOFF, H.L.: Automorphic forms of singular weight are singular forms. Math. Ann. 215, 173-193 (1975).

[50] RESNIKOFF, H.L.: Theta Functions for Jordan Algebras. Inventiones Math. 31, 87-104 (1975).

[51] SATAKE, I.: On the compactification of the Siegel space. J. Indian Math. Soc. 20, 259-281 (1956).

[52] SERRE, J.-P.: A Course in Arithmetic. Graduate Texts in Mathematics 7. New York-Heidelberg-Berlin: Springer (1973).

[53] SIEGEL, C.L.: Über die analytische Theorie der quadratischen Formen. Ann. of Math., II. Ser., 36, 527-606 (1935); Ges. Abh.I, 326-405.

[54] SIEGEL, C.L.: Einführung in die Theorie der Modulfunktionen n-ten Grades. Math. Ann. 116, 617-657 (1939); Ges. Abh.II, 97-137.

[55] SIEGEL, C.L.: Symplectic Geometry. Amer. J. Math. 65, 1-86 (1943); Ges. Abh. II, 274-359.

[56] SIEGEL, C.L.: Die Modulgruppe in einer einfachen involutorischen Algebra. Festschrift Akad. Wiss. Göttingen 1951, 157-167; Ges. Abh. III, 143-153.

[57] SIEGEL, C.L.: Über die algebraische Abhängigkeit von Modulfunktionen n-ten Grades. Nachr. Akad. Wiss. Göttingen 1960, 257-272; Ges. Abh. III, 350-365.

[58] STEIN, E.M., and WEISS, G.: Introduction to Fourier Analysis on Euclidean Spaces. Princeton: Princeton University Press (1971).

[59] VIGNÉRAS, M.-F.: Arithmétique des Algèbres de Quaternions. Lecture Notes in Mathematics 800. Berlin-Heidelberg-New York: Springer (1980).

[60] VOLKMANN, U.: Thetafunktionen. Diplomarbeit, Münster (1982).

[61] WEYL, H.: Theory of reduction for arithmetical equivalence I. Trans. Amer. Math. Soc. 48, 126-164 (1940). II. Trans. Amer. Math. Soc. 51, 203-231 (1942).

[62] WITT, E.: Eine Identität zwischen Modulformen zweiten Grades. Abh. Math. Sem. Hansische Univ. 14, 323-337 (1941).

List of symbols

$\bar{a}$	2
$\hat{a}$, $\check{a}$	14
$a \mid_l b$, $a \mid_r b$	10
$a \parallel b$	11
A[B]	21
$\mathrm{Aut}(S;\mathbb{F})$	38
$\mathrm{Bih}\ \mathcal{U}$	48
$\mathrm{Bih}\ H(n;\mathbb{F})$	50
$C(m,n;\mathbb{F})$	146
$C(n;\mathbb{F})$	58,146
$c_n(N;\mathbb{F})$	37
$(\det M\{Z\})^k$	77,78
det X , det Z	21,47
diag S	26
dS	40
$d\nu$	51,52
$e_1,\dots,e_r$	2
$E(\mathbb{F})$	4
$E(n;\mathbb{F})[\alpha]$	33
$E^k_{n,j}(Z,f)$	152
$E^k_n(Z)$	155
$\mathbb{F} = \mathbb{R},\mathbb{C},\mathbb{H}$	2
(f,g)	166
$F(C)$	69
$F(\mathbb{F})$, $F^*(\mathbb{F})$	6
$F(n;\mathbb{F})$	58
$F_{n,j}$	146
$f \mid_k M$	78
gcrd	10
$GL(n;\mathbb{F})$	14
$GL(n;\mathcal{O})$	15,16
$h = h(n;\mathbb{F})$	20
$H(n;\mathbb{F})$	46
I_{jk}	16
Im Z	46
$I(\mathcal{O})$	7
J	43
$\mathrm{Mat}(m,n;\mathbb{F})$, $\mathrm{Mat}(n;\mathbb{F})$	14
$\mathrm{Mat}(m,n;\mathcal{O})$, $\mathrm{Mat}(n;\mathcal{O})$	15
$\mathrm{Mat}(n;\mathbb{F})_{\mathbb{C}}$	46
$M^k_{n,j}$	167
$M_n = M(\Gamma(n;\mathbb{F}))$	183
$M^s(\Gamma(n;\mathbb{F}))$	190
$M_1 \times M_2$	44
$M\langle Z\rangle$, $M\{Z\}$	49
$M\{Z,W\}$	157
N(a)	2
$\mathcal{O}(\mathbb{F})$, $\mathcal{O}^{\#}(\mathbb{F})$	4,12
$P^k_{n,j}(Z,W,\varphi)$	159
$P^k_{n,j}(Z,W)$	161
$P^k_n(Z,\varphi)$	175
$\mathrm{Pos}(n;\mathbb{F})$	23
$P \equiv Q \bmod I$	67,118
$Q^k_{n,j}(Z,T)$	162
$r = r(\mathbb{F})$	2
Re(a)	2
Re Z	46
$R(n;\mathbb{F})$	29
$s = s(\mathbb{F})$, $S(\mathbb{F})$	5
$S_{\mathbb{F}}$	113,114
$s(n;\mathbb{F})$	66
$S(n;\mathbb{F})[\alpha]$	63
$\mathrm{Sp}(n;\mathbb{F})$	43
$\mathrm{st}(m;\mathbb{F})$	115
$\mathrm{Sym}(n;\mathbb{F})$	20
$\mathrm{Sym}(n;\mathcal{O})$	54
$\mathrm{Sym}^{\tau}(n;\mathcal{O})$	72
tr(X)	20
$\nu(n;\mathbb{F})$	40

$\mathrm{vol}(L)$	73
$\mathrm{vol}(\Gamma(n;\mathbb{F}))$	64
$V_{\varepsilon}(n;\mathbb{F})$	148
$\hat{X}$, $\check{X}$, $\bar{X}'$	15,20,23
$\hat{Z}$, $\check{Z}$, $\tilde{Z}$, $\bar{Z}'$	46
$\Gamma_n = \Gamma(n;\mathbb{F})$	54
$\Gamma(n;\mathbb{F})[q]$	67
$\Gamma(n;\mathbb{F})\langle q\rangle$	67,120
$\Gamma_{n,j}$	146
$[\Gamma(n;\mathbb{F}),k]$	78
$[\Gamma(n;\mathbb{F}),k]_o$	83
$[\Gamma(n;\mathbb{F}),k]^s$	96
$[\Gamma(n;\mathbb{F}),k]_{\Theta}$	115
ϕ	83
Λ , $\Lambda(\mathbb{F})$, Λ^{τ}	101,103,105
$\mu(S;\mathbb{F})$	28
$\Theta_{\mathbf{P},Q}(Z,S;\Lambda)$	101
$\Theta(Z,S;\mathbb{F})$	113
Σ_n , $\Sigma_{n,j}$	145
$\tau(A,B)$	20,72,100
$\#(S,T;\mathbb{F})$	38
$>$, $\gtrsim$	23,26

Index

algebraically dependent, - independent 89,90
analytic class invariant 75
automorph 38

biholomorphic 47

character 92
class number 37
completion of squares 21
congruence subgroup 67
congruent modulo *I* 67,118
cusp form 83

degree 46,78
divisor, left -, right -, total - 10,11

EISENSTEIN-series 152,155
Elementary divisor theorem 16
elementary set 33
even, $\mathbb{F}$-even 107

FOURIER-expansion 73,77,79,93
fundamental domain 6,35,65,69
fundamental relations 43

group of units 4

HADAMARD's inequality 24
half-space 46
HERMITE's inequality 29
Hermitian matrix 20
holomorphic 47,101

imaginary part 46
integrally equivalent 28
integral matrix 15

invariant element 7

JACOBIan representation 25

KOECHER's principle 76

lattice, dual lattice 3,12
level of a congruence subgroup 67
level of a matrix 119

Main theorem of reduction theory 35
meromorphic 183
minimum 28
MINKOWSKI's conditions of reduction 29
modular form 78,92
modular function 183,189
modular group 54
modular transformation 54

number of representations 38

ordering 3

POINCARÉ-series 159,162
positive definite, - semi-definite 23
principle congruence subgroup 67

quaternions, - of HURWITZ, even - 2,4,13

real part 2,46
reduced matrix 29

scalar product 166
SIEGEL's elementary set 63

SIEGEL's ϕ-operator 83
singular modular form 126
stable, $\mathbb{F}$-stable 113
stable modular form 129
symmetric modular form 96
symplectic group 43
symplectic transformation 50
symplectic volume element 53

theta-group 67
theta-series 101
Theta-transformation-formula 111
trace 20

unimodular group 15

vertical strip 148

weight 78,92